The Three Books

Book One: The Last Keeper

Who Watches the Light? On the Human Cost of Trusting Machines — and the Price of Removing the Person from the Loop

A lighthouse keeper who held his post beside a dead man. A radiation therapy machine whose software punished the most competent operators. An aircraft whose pilots received seventy-five stall warnings in fewer than four minutes. A sentencing algorithm that could not be examined by the defendant it condemned. A social media platform whose recommendation engine played a determining role in genocide. Twelve chapters. Five patterns. A century of evidence that removing the human from the loop produces catastrophe at the margin — and that the people who bear the cost are never the people who decided to remove the keeper.

Book Two: Who Holds the Jar?

The Machine Evolves: A Natural History of Computation — and a Warning

Eight thousand years ago, a shepherd pressed a clay token into a jar. The token represented a sheep. The jar was the first ledger. The ledger was the first computer. This book traces what happened next — across fourteen chapters and five millennia, from counting stones through clockwork calculators, vacuum tubes, transistors, microprocessors, the internet, neural networks, quantum circuits, and biological substrates. In every era, the same pattern repeats: computational power distributes, then reconcentrates. Each cycle appears to democratize. Each cycle produces reconcentration. And the current cycle — AI, quantum computing, biological computation — may produce a concentration of power that, for the first time in history, cannot be reversed.

Book Three: The Great Convergence

On the Battle for Who Writes the Rules — and What Happens When Nobody Does

Every transformative technology passes through a governance window — a finite period during which the rules are written. The printing press took 350 years. Electricity took 53. Nuclear energy took 23. The internet was never adequately governed. This book traces the five-phase cycle across four centuries and applies it to AI: the chaos of ungoverned code, the epistemic crisis of synthetic information, the displacement of human judgment, the concentration of infrastructure in seven companies and one satellite constellation, and the geopolitical race between two superpowers with incompatible visions of governance. The window for AI is open. It is measured in years, not decades. And the room where the rules should be written has not yet convened.

The Urgency

This trilogy documents evidence that is accumulating faster than any institution can process.

AI systems are already making decisions that affect the liberty of defendants, the health of patients, the employment of millions, the information environment of billions, the targeting of military operations, and the surveillance of entire populations. These systems are deployed without adequate testing, without transparency, without independent audit, and without the human oversight that a century of evidence shows is irreplaceable when the system encounters conditions it was not designed for.

The infrastructure that powers these systems is concentrated in seven organizations that can train frontier models, two companies that manufacture the required chips, three cloud providers that host the compute, and one private satellite constellation that controls over sixty percent of active satellites in low Earth orbit — a constellation whose owner's decisions have already determined the outcome of military operations in a major European war.

The governance that could constrain this concentration, preserve human oversight, and ensure that the people affected by these systems have a voice in how they are governed does not exist in a form adequate to the challenge. The EU AI Act covers one continent. The United States has not passed comprehensive AI legislation. No international framework exists. The room has not convened.

The Alternative

The alternative to paying attention now is not that everything continues as before. The alternative is that the rules get written without you.

Every governance framework in history has reflected the interests of whoever was in the room when it was written. The Constitution reflected the interests of propertied white men and required two centuries of amendment. The internet's terms of service — the rules that govern the speech, commerce, and associations of billions of people — were written by corporate lawyers in Mountain View and Menlo Park. The alternative to governance is not freedom. It is rule by whoever fills the vacuum.

The printing press's ungoverned chaos phase lasted 130 years and killed millions. The internet's ungoverned chaos produced surveillance capitalism, platform monopolies, and algorithmic amplification of genocide. The alternative to governing AI is not that AI goes ungoverned. It is that AI is governed by the companies that build it, on terms they write, for purposes they determine, without democratic mandate and without accountability to the people who live under the rules.

"The alternative to governance is not freedom. It is rule by whoever fills the vacuum. And the vacuum is being filled right now, by the people who benefit most from the absence of rules."

Why These Books

There are many books about AI. Most are written by the people who build it or the people who sell it. This trilogy is written for the people who live under it — the patients, the defendants, the workers, the citizens, the voters, the parents, the people whose data trains the systems and whose lives are shaped by the decisions those systems make.

The Last Keeper will make you see the cost. Every chapter opens with a specific person in a specific moment — the pilot at two in the morning, the patient on the table, the defendant in the courtroom — and traces the architecture of failure that put them there. You will not read about AI the same way afterward.

Who Holds the Jar? will make you see the pattern. Five thousand years of computational history, told as a single narrative, revealing a cycle so persistent that recognizing it is the only defense against repeating it. You will not look at the current AI concentration the same way afterward.

The Great Convergence will make you see the window. The governance window is open. The evidence is documented. The question is whether the room convenes before it closes — and whether the people who bear the consequences have a seat at the table when it does. You will not accept "it's too early to regulate" the same way afterward.

Three books. Thirty-eight chapters. 144 references
From a clay jar in Mesopotamia to drone submarines in the Black Sea.
From a lighthouse keeper beside a dead man to an AI system
that generates military targets in twenty seconds.
From a room in Philadelphia that changed history
to the room that has not yet convened for artificial intelligence.

"The choice is not between perfect governance and imperfect governance. It is between imperfect governance and none. And if the history this trilogy has documented teaches anything, it is that none is not an absence. It is a choice — made by default, by the people who benefit from the absence, at the expense of everyone else."

Pumulo Sikaneta

The Last Keeper

Who Watches the Light?

The Last Keeper

Who Watches the Light?

On the Human Cost of Trusting Machines —
and the Price of Removing the Person from the Loop

Pumulo Sikaneta

For the keepers —
the ones who stayed at the light
when the mechanism said they were no longer needed.

Contents

Author's Preface

This book began with a lighthouse.

In 1801, a man named Thomas Howell kept a light burning for weeks on a rock in the Atlantic, beside the decomposing body of his dead companion, because the light was not for him. The light was for the ships. The ships could not see the rocks. The ships depended on a human being who sat in the dark and kept the mechanism working when the mechanism alone was not enough.

I could not stop thinking about Thomas Howell after I first encountered his story. Not because of the heroism — though it is extraordinary — but because of what happened next. The mechanism improved. The light automated. And the keeper was removed. Not because the keeper had failed but because the machine had become reliable enough that the keeper seemed unnecessary.

The same pattern, I discovered, repeated in every domain I examined. Radiation therapy machines that killed patients because the safety interlocks were moved from hardware to software. Aircraft that crashed because the autopilot was so reliable that the pilots forgot how to fly. Sentencing algorithms that encoded racial bias and called it objectivity. Social media platforms that optimized for engagement and amplified genocide. Military targeting systems that compressed human review to a formality. In every case: a system that worked on average, a keeper who was removed or degraded, and a catastrophe that arrived at the margin.

This book documents that pattern. It is not a book against technology. It is a book about the specific, recurring, documentable cost of removing the human from the loop — the person who exercises judgment when the system encounters conditions it was not designed for. The evidence spans two centuries and a dozen domains. The argument is simple: the keeper is not a cost. The keeper is the thing that stands between the system's average performance and the catastrophe at the margin. And the margin is where people live.

This book owes a particular debt to my brother, Dr. Tabo Sikaneta, a nephrologist at the Scarborough Health Network in Toronto and founding member of the African Caribbean Kidney Association. Chapter 7 documents the Obermeyer study — a US healthcare algorithm that used spending as a proxy for illness, systematically disadvantaging Black patients because they spent less in a system where spending correlates with access. One might conclude that the problem is specific to the American system: remove the cost barrier with universal healthcare and the disparity disappears.

Dr. Sikaneta's 2025 *BMJ Open* study demonstrates that it does not. His research in Scarborough — Toronto's most diverse region, home to one of North America's largest dialysis programs — found that dialysis prevalence was 4.2 times higher among immigrants than non-immigrants within Canada's universal healthcare system. But the finding that matters most for this trilogy is what the data showed beneath that headline: the disparity was not uniform across all immigrants. It was concentrated in patients born in the Caribbean, Southeast Asia, and South Asia. And critically, the year of immigration did not influence the outcomes. A patient who arrived from the Caribbean thirty years ago showed the same trajectory as one who arrived five years ago. The disparity is not an acculturation problem. It is not a "new immigrant" problem. It is something embedded deeper — in genetics, in epigenetics, in environmental exposures carried from the country of origin, in structural factors that persist across decades in a new system with universal access.

This raises the question that Chapter 7 cannot answer but must ask: what are the right data points to use, and how do we create universal measurements in a field as complex as medicine, where the categories themselves — race, ethnicity, birth country, immigration status — may each capture a different fragment of the underlying reality? An algorithm trained on Canadian data would not have the Obermeyer cost-need confusion. But it would still encode the disparities Dr. Sikaneta's research documents — because the disparities are in the outcomes, not the billing. We do not always have the clinical expertise and the shrewd perception of a Dr. Sikaneta to look beneath the numbers and ask whether the data categories themselves are adequate to the complexity they claim to represent. And in healthcare, the cost of using the wrong dataset, or the wrong proxy, or the wrong category is not an error rate. It is a patient. Dr. Tabo Sikaneta is a keeper in the truest sense this book describes — the person who looks at what the system is measuring and asks whether it is measuring the right thing. I am proud that he is my brother. I am prouder that he is his patients' keeper.

Pumulo Sikaneta

The Last Keeper

The Keeper

On a dead man and a burning light, the vigil that could not be automated, and the question that connects a rock in the Irish Sea to every machine you will ever trust with your life

> *"The keeper is not a redundancy. The keeper is the judgment. Remove the judgment and you have a system that works perfectly right up until the moment it doesn't — and then there is nothing between the failure and the consequence."*

The Keeper

On a dead man and a burning light, the vigil that could not be automated, and the question that connects a rock in the Irish Sea to every machine you will ever trust with your life

Smalls Rock sits in the Irish Sea, roughly twenty miles off the coast of Pembrokeshire, Wales. It is barely a rock at all — a low, jagged outcropping that disappears beneath the waves at high tide, invisible to any ship that does not already know it is there. For centuries, it killed sailors. Ships struck the rock in darkness, in fog, in storms, and went down with all hands. The sea around Smalls was a graveyard before anyone thought to put a light on it.

In 1776, a lighthouse was built on the rock — an extraordinary feat of engineering for the era, a wooden structure on iron stilts driven into the stone, exposed to the full violence of the Atlantic. The lighthouse required two keepers at all times. They lived on the rock, maintained the light, and endured conditions that most people today would find difficult to imagine: weeks of isolation, storms that shook the structure to its foundations, waves that broke over the lantern room seventy feet above the waterline. The keepers were not comfortable. They were not well paid. What they were was necessary. Without them, the light went dark. And when the light went dark, people died.

In the winter of 1801, one of the two keepers at Smalls — Thomas Griffith — died. The circumstances of his death are not precisely recorded, but what followed is. His partner, Thomas Howell, was alone on the rock with the body. A storm was raging. No boat could reach the lighthouse. Howell could not bury the body at sea, because he feared that if the body disappeared, he would be accused of murder when he was eventually relieved. So he built a crude coffin from salvaged materials and lashed it to the exterior railing of the lighthouse.

The storm persisted for weeks. The coffin broke apart. The body, exposed to the wind and salt, decomposed in view of the lantern room. And Thomas Howell kept the light burning.

He kept it burning for the entire duration of his isolation — some accounts say weeks, others say months, until a relief crew could finally reach the rock. When they arrived, Howell was emaciated, traumatized, and, according to contemporary reports, so changed by the ordeal that his friends did not recognize him. But the light had not gone dark. Not for a single night. Because Howell understood something that no mechanism could understand: the light was not for him. The light was for the ships. And the ships could not know whether the keeper was suffering, or sane, or alone with a dead man in a storm. They could only know whether the light was on.

> *"Thomas Howell kept the light burning for weeks alongside a decomposing body in an Atlantic storm, because he understood something no mechanism could: the light was not for him. The light was for the ships."*

What a Keeper Does

I begin with this story not because it is pleasant but because it is precise. It isolates, with a clarity that more complex examples cannot match, the thing that is lost when the human is removed from a system.

A lighthouse mechanism can keep a flame burning. A clockwork assembly can rotate a lens. A timer can switch a light on at dusk and off at dawn. These are tasks that machines perform more reliably than humans — they do not fall asleep, do not forget, do not get distracted. On the average night, the machine is better than the keeper.

But Thomas Howell did not do what a machine does. He did not merely maintain a flame. He made a judgment — a continuous, sustained, morally weighted judgment — that the light must remain on regardless of his own circumstances. He evaluated a situation that no designer could have anticipated (a dead partner, a broken coffin, a weeks-long storm), and he chose to keep the light burning despite every human reason to abandon the post. A mechanism does not make this choice, because a mechanism does not encounter situations it was not designed for. It encounters the conditions it was built to handle, or it fails. There is nothing in between.

The keeper occupies the in-between. The keeper is the layer of judgment that operates in the space between what the system was designed for and what actually happens. The system handles the expected. The keeper handles the rest. And "the rest" is where people live and die.

This book is about what happens when the keeper is removed.

The Removal

Lighthouses were automated throughout the twentieth century. The last manned lighthouse in England was automated in 1998. The last in the United States was Boston Light, automated in 1998. The last keeper in Scotland left Flannan Isles in 1971. Country by country, rock by rock, the keepers departed and the machines took over.

The case for automation was overwhelming and, on its own terms, correct. Automated lights are more reliable than human-maintained lights. They do not require expensive support logistics — supply ships, fresh water, food, relief crews. They do not expose human beings to the brutal conditions of offshore rocks. They do not create the psychological toll documented in generations of keepers: depression, alcoholism, the slow erosion of sanity in isolation. The cost per year of operation dropped by an order of magnitude. The failure rate dropped as well. On every measurable metric, automation was superior.

Yet the automated systems could not see a ship in distress and radio for help. The keepers could and did — lighthouse keepers were responsible for hundreds of documented rescues over the centuries, pulling survivors from wrecks, launching lifeboats in conditions no sane person would enter, and providing shelter to shipwrecked sailors who reached the rock alive. The automated systems could not observe unusual weather patterns and warn approaching vessels by signal. They could not detect a malfunction in the light that was subtle enough to pass the mechanism's self-check but visible to a human eye — a lens slightly fouled, a rotation slightly irregular, a beam whose pattern had shifted just enough to confuse rather than guide.

The automation was superior on every measurable metric. The keeper was superior on the metrics that could not be measured — because the situations that required judgment were, by definition, the situations that were not anticipated, and you cannot measure the value of handling a situation you did not know would occur.

The Pattern

I am going to trace a pattern through this book that I believe is among the most important and least discussed dynamics in the history of technology. The pattern is this:

A system is built. The system includes a human in the loop — a person whose job is to exercise judgment when the system encounters conditions it was not designed for. The system works. It works well. It works so well that the human's contribution becomes invisible, because the human is only needed in the rare cases when the system fails, and the system rarely fails. Someone calculates the cost of the human. Someone compares it to the cost of the system operating without the human. The human is expensive. The system is cheap. The human is removed.

For a while — sometimes years, sometimes decades — nothing bad happens. The system operates without the human and performs as well as it did with the human, because the conditions remain within the range the system was designed for. The removal is declared a success. The savings are celebrated. The human is forgotten.

Then the system encounters a situation it was not designed for. And there is no one there.

The consequence is proportional to the stakes. In a lighthouse, a ship may go unwarned. In an aircraft cockpit, 228 people die over the Atlantic. In a radiation therapy room, a patient receives a hundred times the intended dose. In a criminal courtroom, an algorithm sentences a person based on the color of their neighborhood. In a military targeting system, a weapon selects a civilian structure because the pattern matched a probability threshold and no human was empowered to say: wait.

This is not a book about machines failing. All machines fail. Failure is an engineering problem, and engineering problems have engineering solutions. This is a book about the systematic removal of the human who could catch the failure — and the specific, documented, often fatal consequences that follow.

What I Will Show

Across twelve chapters, I will trace this dynamic through five domains: the systems that move us, the systems that heal us, the systems that judge us, the systems that watch us, and the systems that fight for us. In each domain, the same five patterns recur.

The **Reliability Trap**: the more reliable the automated system, the less practiced the human becomes at the task, and the worse the human performs when the system fails. The autopilot that flies perfectly for a thousand hours produces a pilot who cannot fly the thousand-and-first.

The **Single Point of Trust**: catastrophic failures trace, with remarkable consistency, to a system that trusted one sensor, one algorithm, or one assumption — without independent verification. One data source. One chance to be wrong. One failure away from disaster.

The **Invisible Bias**: when human decisions are encoded into automated systems, the biases travel with them — but stripped of the human context that might have recognized and corrected them. The bias becomes mathematics. Mathematics does not look biased. It looks objective. That is what makes it dangerous.

The **Speed-Judgment Tradeoff**: automated systems operate faster than human thought. As systems accelerate, the window for human intervention shrinks. At sufficient speed, the human is no longer in the loop. The human is beside it — a spectator to decisions that have already been made.

The **Accountability Gap**: when a human makes a catastrophic error, the chain of responsibility is clear. When an automated system does, responsibility diffuses — across designers, operators, manufacturers, regulators, and the system itself. Often, nobody is held accountable. The error is attributed to the system. The system is not a person. The consequence falls on the victim. The lesson is not learned.

Why This Book Exists

I wrote this book because the pattern would not leave me alone. The same architecture of failure — the keeper removed, the system trusted, the catastrophe arriving at the margin — repeated in every domain I examined: aviation, medicine, criminal justice, technology, warfare. The question that unifies every chapter is the one that makes this pattern matter to every individual: when the machine is wrong — and it will be wrong, not once but repeatedly, not in some distant future but in the systems you depend on right now — will there be a human present, empowered, and skilled enough to catch the failure before it costs a life?

This book is about people. Specific people in specific moments: a pilot over the Atlantic at two in the morning with an autopilot that has just disconnected. A patient on a table beneath a radiation beam that the software has miscalibrated. A defendant in a courtroom facing a sentence that an algorithm calculated from data the defendant cannot see or challenge. A family in a building that an autonomous targeting system has identified as a military objective.

The people in these stories did not consent to the removal of the keeper. Most did not know the keeper had been removed. They trusted the system because the system had been reliable, because the institution behind the system was credible, and because nobody told them that the human who used to stand between them and the machine's errors was no longer there.

Thomas Howell kept the light burning because he understood that the ships could not see whether he was suffering. The ships could only see the light. This book is for the ships — for the people who depend on systems they cannot see inside, maintained by keepers they do not know have been removed.

> *"This book is for the ships — for the people who depend on systems they cannot see inside, maintained by keepers they do not know have been removed."*

The Question

The automated lighthouse at Smalls still operates. The light is powered by solar panels and batteries. The lens rotates on electric motors. The system is monitored remotely from the mainland. It has been decades since a keeper stood on the rock.

The light is more reliable now than Thomas Howell could have maintained it. The point is that there will be nights when the mechanism fails. Not a night. Nights. There will be storms the designers did not model, corrosion the sensors did not detect, conditions that fall outside the parameters the system was built to handle. The failures will not announce themselves. They will arrive quietly, in the gap between what the system was designed for and what actually happens, and they will keep arriving for as long as the system operates. Because the gap is permanent. No system, however sophisticated, can anticipate every condition it will face. That is not a limitation of engineering. It is a property of the world.

The question this book asks is not whether the machines will fail. They will, and they will fail more than once, and the failures will come at the moments when the stakes are highest, because those are the moments that test the boundaries of what the system was designed for. The question is whether, when those nights come — and they will keep coming, for as long as we build machines and trust them with consequential tasks — there will be a keeper watching the light.

The Last Keeper

The Unsinkable Ship

On the RMS Titanic, the belief that sufficient engineering eliminates
the need for caution, and the oldest form of the reliability trap —
which did not require a computer, only a human who trusted one
machine too much

"Not even God himself could sink this ship."
— attributed to a Titanic crew member, reported by survivor Sylvia Caldwell

The Unsinkable Ship

On the Titanic, the belief that sufficient engineering eliminates the need for caution, and the oldest form of the reliability trap

Everyone knows the Titanic struck an iceberg. Almost everyone misunderstands why.

The standard telling is an engineering story: the ship was poorly designed, the steel was brittle, the rivets were weak, the watertight compartments were not watertight enough. These details are largely accurate and entirely beside the point. The Titanic did not sink because the engineering failed. The engineering performed exactly as it was designed to perform. The hull withstood the impact along its length. The watertight doors closed automatically. The pumps activated. The systems worked. The ship still sank, because the humans who operated it had made a series of decisions, hours and days before the collision, that the best engineering in the world could not compensate for.

The Titanic sank because the people responsible for it trusted the machine more than they trusted the warnings.

The Warnings

On April 14, 1912 — the day of the collision — the Titanic received at least six ice warnings from other ships. The messages came through the ship's Marconi wireless room, a small cabin staffed by two radio operators: Jack Phillips and Harold Bride. The operators were employed not by the White Star Line but by the Marconi Company, and their primary responsibility was handling passenger telegrams — wealthy passengers sending and receiving personal messages from the middle of the Atlantic. The ice warnings were incidental to their main workload.

Some of the warnings reached the bridge. Some did not. The one from the SS Mesaba, received at approximately 9:40 PM — less than two hours before the collision — described a large field of ice directly in the Titanic's path. It was not delivered to the bridge because Phillips was busy clearing a backlog of passenger telegrams. He acknowledged the message and placed it under a paperweight on his desk. It was found there after the sinking.

Here is the first keeper failure, and it has nothing to do with engineering. The information existed. The warning was received. It was specific, timely, and accurate. But the system — the organizational system, not the ship — did not treat ice warnings as a priority. They were one data stream among many, handled by operators whose primary incentive was passenger satisfaction, not navigation safety. The keeper was present. The keeper received the warning. The keeper put it under a paperweight because the system he operated in did not empower him to act on it.

The Speed

The Titanic was traveling at approximately 22.5 knots — nearly its maximum speed — through waters known to contain ice. This was not an oversight. Captain Edward Smith was aware of the ice warnings that had reached the bridge. He adjusted the ship's course slightly southward. He did not reduce speed.

The decision not to reduce speed was consistent with standard maritime practice of the era and with the institutional culture of the White Star Line. Slowing down meant arriving late. Arriving late meant disappointing passengers and the press. The Titanic's maiden voyage was a commercial event as much as a nautical one, and the incentive structure — the pressure from the company, the expectations of the passengers, the competition with Cunard Line for the North Atlantic record — favored speed over caution.

Captain Smith had thirty-eight years of experience at sea. He was the most senior captain in the White Star fleet. He had navigated ice before, many times, without incident. His experience told him that ice could be spotted in time to steer around it, that the lookouts would do their job, that the ship's size and construction provided a margin of safety that smaller vessels did not enjoy. His experience was extensive, and it was wrong.

This is the mechanism I want to name, because it will recur throughout this book. It is not recklessness. Captain Smith was not reckless. He was experienced, competent, and operating within the norms of his profession. What he was, specifically, was *calibrated to a system that had never failed*. He had crossed the Atlantic with ice warnings before. He had maintained speed before. Nothing bad had happened before. His judgment was formed by a track record of success, and that track record of success produced a confidence that the next crossing would be like the last. The keeper was present, experienced, and empowered. The keeper's judgment was shaped by a machine that had never failed him — and so the keeper trusted the machine more than he trusted the warning.

> *"Captain Smith was not reckless. He was calibrated to a system that had never failed. His judgment was formed by a track record of success. And a track record of success produces a confidence that the next time will be like the last."*

The Boats

The most telling detail about the Titanic disaster is not the collision. It is the lifeboats.

The Titanic carried twenty lifeboats with a total capacity of 1,178 people. There were 2,224 people aboard. The ship could carry roughly half its passengers and crew to safety. The other half had no provision.

This was not an accident and it was not illegal. The British Board of Trade regulations governing lifeboat capacity had been written in 1894, when the largest ships were under 10,000 gross tons. The Titanic was 46,328 gross tons. The regulations had not been updated. Under the rules as they stood, the Titanic actually exceeded the minimum lifeboat requirement. It was, by the standards of its regulatory framework, overprovided with lifeboats.

But the regulatory failure is only part of the story. The deeper truth is that the White Star Line's original plans for the Titanic included forty-eight lifeboats — enough for everyone aboard. Alexander Carlisle, the managing director of the shipyard that built the Titanic, designed the davit system to accommodate this number. The decision to carry twenty instead of forty-eight was made by the White Star Line's managing director, J. Bruce Ismay, who judged that the additional boats would clutter the deck and alarm passengers. The reasoning was explicit: a ship this well-engineered did not need that many lifeboats. The lifeboats were a concession to regulation, not a reflection of genuine risk. They were carried because the law required them, not because anyone expected to use them.

This is the reliability trap in its purest pre-digital form. The engineering was so good that the safety measure was perceived as unnecessary. The backup was removed — not physically, but functionally, by reducing it below the level that would actually serve its purpose — because the primary system was trusted to make the backup irrelevant.

The Night

At 11:40 PM on April 14, 1912, lookout Frederick Fleet spotted the iceberg from the crow's nest and struck the warning bell three times. He telephoned the bridge: "Iceberg, right ahead." First Officer William Murdoch ordered the helm hard to starboard and the engines reversed. The ship began to turn.

It was not enough. The Titanic struck the iceberg along its starboard side below the waterline, opening the hull across approximately 250 feet. Water flooded five of the sixteen watertight compartments. The ship was designed to remain afloat with four compartments flooded. Five was one too many. Thomas Andrews, the ship's designer, who was aboard for the maiden voyage, calculated that the ship would sink in approximately one to two hours. He was correct.

The lookouts, it should be noted, were operating without binoculars. The ship's binoculars had been locked in a cabinet, and the officer who had the key had been reassigned before departure and had taken the key with him. Nobody had replaced the binoculars or forced the cabinet open. On a calm, moonless night — the worst conditions for spotting ice, because there are no waves to break against the berg and no moonlight to illuminate it — the lookouts were scanning the horizon with their naked eyes.

The keeper was in the crow's nest. The keeper was doing his job. The keeper was denied the tool that would have given him the seconds of additional warning that might have made the difference — because the tool was locked in a cabinet and nobody had thought it mattered enough to fix.

The Pattern, Named

The standard Titanic narrative is a morality tale about hubris — arrogant men who built an unsinkable ship and were punished by nature for their arrogance. That narrative is satisfying and incomplete. The Titanic was not brought down by arrogance in the theatrical sense. It was brought down by a perfectly rational, perfectly understandable, and perfectly lethal form of trust.

The engineering was trusted, so the lifeboats were reduced. The ship's track record was trusted, so the speed was maintained. The lookout system was trusted, so the missing binoculars were not replaced. The wireless system was trusted, so the ice warnings were treated as low priority. At every decision point, the people involved were not ignoring the risk. They were evaluating the risk against the reliability of the system, and concluding — reasonably, based on experience — that the system would handle it.

The system handled it right up until the night it didn't. And 1,517 people died because the preparations for that night had been made by people who did not believe it would come.

I said in the previous chapter that this book is about the systematic removal of the human who could catch the failure. The Titanic is the foundation stone of that argument, because it demonstrates that the pattern does not require a computer. It does not require artificial intelligence. It does not require any technology more sophisticated than a steam engine and a wireless telegraph. It requires only a system that works well enough, for long enough, that the humans who depend on it stop preparing for the night it doesn't.

"The Titanic was not brought down by arrogance. It was brought down by a perfectly rational, perfectly understandable, and perfectly lethal form of trust. The people involved were not ignoring the risk. They were evaluating it against a system that had never failed."

The Bridge

The Titanic went down in 1912. The first electronic computer was built in 1944. Between these two dates, the pattern I have described — the reliability trap, the single point of trust, the keeper present but not empowered — repeated itself through industrial disasters, aviation accidents, medical errors, and military miscalculations. The details varied. The structure did not.

What changed with computers — and what makes the next ten chapters different from this one — is scale and speed. The Titanic's officers had hours to process ice warnings. A modern autopilot disconnects in milliseconds. The Titanic's lifeboat capacity was a decision made in a boardroom over weeks. An algorithm sentences a defendant in seconds. The human failures aboard the Titanic were failures of judgment, operating at human speed, with human time to reconsider. The failures in the chapters that follow are the same failures of judgment, encoded into machines that operate at a speed that leaves no room for reconsideration.

Computers did not invent the reliability trap. They automated it. And an automated trap springs faster than a human can react.

The Last Keeper

The Invisible Interlock

On the Therac-25 radiation therapy machine, the software that replaced a physical safety mechanism, and the patients who died because a timing error in the code could only be triggered by an operator who was good at her job

"The most dangerous phrase in the language is 'we've always done it this way.' The second most dangerous is 'the software handles that.'"

The Invisible Interlock

On the Therac-25 radiation therapy machine, the software that replaced a physical safety mechanism, and the patients who died because a timing error could only be triggered by an operator who was good at her job

A radiation therapy machine is, in principle, a straightforward device. It generates a beam of high-energy particles — electrons or X-rays — and directs that beam at a tumor inside a patient's body. The beam destroys the tumor cells. The skill is in the precision: the beam must be strong enough to kill the tumor and narrow enough to spare the healthy tissue surrounding it. Too little radiation and the tumor survives. Too much and you destroy the patient along with the disease.

The difference between a therapeutic dose and a lethal dose is not a matter of kind. It is a matter of degree. The same machine that saves a life at 200 rads will end one at 20,000. The margin between healing and killing is a number on a screen, set by an operator, controlled by software, and delivered by a machine that the patient cannot see, cannot understand, and has no choice but to trust.

Between June 1985 and January 1987, a radiation therapy machine called the Therac-25 delivered massive overdoses to at least six patients at clinics in the United States and Canada. The overdoses ranged from approximately 13,000 to 25,000 rads — roughly one hundred times the intended dose. Three patients died from the radiation injuries. The others suffered severe and permanent harm: radiation burns that penetrated deep into tissue, neurological damage, and the slow, agonizing progression of radiation sickness in organs that had received doses designed for industrial sterilization, not human therapy.

The machine's operator console displayed the message "MALFUNCTION 54" after the overdoses. The operators did not know what Malfunction 54 meant. The manual did not explain it. The manufacturer's technical support could not immediately identify it. In at least one case, an operator, seeing the malfunction message but believing the machine had simply failed to deliver the dose, reactivated the beam and irradiated the patient a second time.

"The difference between a therapeutic dose and a lethal dose is not a matter of kind. It is a matter of degree. The margin between healing and killing is a number on a screen."

The Predecessor

To understand what went wrong with the Therac-25, you must first understand what went right with the Therac-20.

The Therac-20 was the previous model in the same product line, manufactured by Atomic Energy of Canada Limited (AECL). It was a dual-mode machine: it could deliver either a low-energy electron beam directly to the patient or a high-energy X-ray beam produced by directing the electrons at a tungsten target. The two modes required different configurations of the machine's internal components — in electron mode, a scanning magnet spread the beam evenly; in X-ray mode, a metal target and a flattening filter shaped the beam for deeper penetration.

The critical point is this: in X-ray mode, the electron beam was enormously powerful — powerful enough to penetrate the tungsten target and produce X-rays. If that full-power beam were delivered directly to a patient without the target in place, the dose would be lethal. The Therac-20 prevented this with **hardware interlocks** — physical mechanisms that made it mechanically impossible for the machine to fire the full-power beam unless the target and flattening filter were correctly positioned. These interlocks operated independently of the software. They were physical barriers. They could not be overridden by a software error, because they were not controlled by software. They were controlled by gravity, springs, and mechanical linkages that either allowed or blocked the beam based on the physical position of the components.

The Therac-20 had software too. The software controlled the treatment parameters, the operator interface, and the sequencing of the machine's operations. But the software did not control the safety-critical function — ensuring that the right components were in place before the beam fired. That function was performed by the hardware, independently, every time, regardless of what the software did.

A door with two locks: a deadbolt and an electronic lock. The deadbolt works by physical mechanism — a metal bar slides into a metal frame. The electronic lock works by software — a computer checks a code and signals a motor. If the electronic lock malfunctions, the deadbolt still holds the door shut. The two systems are independent. The failure of one does not compromise the other. The Therac-20 was a door with two locks.

The Upgrade

The Therac-25 was the next-generation machine. It was faster, more versatile, and more computer-controlled than the Therac-20. It used a newer, more powerful processor. It offered a more sophisticated operator interface. And it removed the hardware interlocks.

The designers reasoned that the software could perform the safety checks that the hardware interlocks had previously provided. The software already controlled the treatment parameters. It already knew which mode the machine was in. It could verify, through software logic, that the correct components were in place before firing the beam. The hardware interlocks were, in this reasoning, redundant — an expensive, mechanical duplication of a function that the software already performed.

The door was given one lock instead of two. The deadbolt was removed because the electronic lock was believed to be sufficient.

This decision was not made carelessly. It was made by engineers who understood the system they were building. The software had been derived, in significant part, from the Therac-20's software, which had operated for years without a safety-critical failure. The reasoning was the same reasoning we saw in the previous chapter: the system has worked. The system has worked for a long time. The backup is expensive and, given the system's track record, unnecessary. Remove the backup.

The Bug

The software had a flaw. It was a type of error called a **race condition** — a situation where the outcome depends on the precise timing of two operations that happen almost simultaneously. To explain this in plain terms, because understanding it is essential to understanding why the patients died.

The Therac-25's operator interface allowed the technician to type in the treatment parameters: the mode (electron or X-ray), the energy level, and the dose. After entering the parameters, the operator could review them on screen and, if a mistake was noticed, move the cursor back to the incorrect field and change it. This was a normal, expected workflow — the kind of correction any operator makes dozens of times a day.

The race condition occurred when the operator changed the mode from X-ray to electron (or vice versa) very quickly — within approximately eight seconds of the initial entry. If the mode was changed fast enough, the software updated the display to show the new mode but did not complete the internal reconfiguration of the machine's physical components. The screen told the operator the machine was in electron mode. The machine was still configured for X-ray mode — with the full-power beam active and no target in place to absorb it.

The beam fired. The patient received the full force of the X-ray-mode electron beam, directly, without the tungsten target, without the flattening filter, at a dose roughly one hundred times higher than intended.

The Cruel Irony

The detail that makes this story structurally important:

The race condition could only be triggered by a fast operator. An operator who typed slowly — who took more than eight seconds to notice a mistake and correct it — would never encounter the bug. The software would have time to complete the internal reconfiguration before the correction was entered. The machine would work correctly.

An experienced operator — someone who had used the machine many times, who was skilled and efficient, who could spot a mistake and correct it in seconds rather than fumbling through the interface — was the only operator who could trigger the lethal error. The bug punished competence. It was invisible to the novice and lethal to the expert.

One of the patients, a man named Ray Cox, was treated at a clinic in Tyler, Texas, in March 1986. The operator entered X-ray mode, then quickly corrected to electron mode. The machine displayed "MALFUNCTION 54." Cox reported feeling a jolt — "like an electric shock," he said — and pain in his shoulder. The operator, believing the machine had simply malfunctioned without delivering the dose, reactivated the beam. Cox received a second massive overdose. He died months later from the radiation injuries.

The operator had no way to know what had happened. The screen said the dose had not been delivered. The manual did not explain what Malfunction 54 meant. The manufacturer had not disclosed that any hazard of this kind was possible. The operator did the reasonable thing — the thing the system's design led her to do. She trusted the screen.

The Investigation

When the accidents were eventually investigated — by the FDA, by independent researchers, and most thoroughly by Nancy Leveson and Clark Turner in what became one of the most cited papers in the history of software engineering — the findings revealed a pattern of failures that extended far beyond the race condition itself.

AECL, the manufacturer, had tested the software. But the tests were designed to verify that the software performed its intended functions correctly — not that it failed safely when something unexpected happened. The race condition was never tested because it arose from an interaction between operator behavior and software timing that the testers did not anticipate. You cannot test for a failure you cannot imagine.

The software had been reused from the Therac-20 — but the context had changed. On the Therac-20, the same software flaw may well have existed. But it didn't matter, because the hardware interlocks would have prevented the beam from firing regardless. The software bug was dormant on the Therac-20 — present in the code but harmless, because the hardware keeper was standing between the bug and the patient. When the hardware keeper was removed in the Therac-25, the dormant bug became lethal.

This is the mechanism this chapter exists to name: **a latent defect in a system that is harmless as long as an independent safety layer exists, and lethal the moment that layer is removed**. The defect does not announce itself when the safety layer is taken away. It waits. It waits for the specific combination of conditions that activates it — and those conditions may not arrive for months or years. The system appears to work perfectly during the interval. The removal of the safety layer is declared a success. And then a patient sits on a table and an operator who is good at her job types a correction in under eight seconds.

The Institutional Failure

After the first reported overdose, at Kennestone Regional Oncology Center in Georgia in June 1985, the hospital's physicist investigated and concluded that the Therac-25 had delivered an overdose. He contacted AECL. The manufacturer investigated and could not reproduce the error. AECL concluded that the machine was safe and resumed operations.

After the second overdose, at the Ontario Cancer Foundation in Hamilton, Ontario, in July 1985, the pattern repeated. The hospital reported the incident. AECL investigated. AECL could not reproduce the error. AECL concluded the machine was safe.

After the third overdose, at the same Hamilton clinic in December 1985, AECL issued a software fix that addressed one potential cause. The fix did not address the race condition, which had not yet been identified.

The fourth, fifth, and sixth overdoses occurred in 1986 and 1987, after the "fix" had been applied. Patients continued to be irradiated at lethal doses. The manufacturer continued to express confidence in the machine's safety. It was not until the FDA intervened, conducted its own investigation, and ordered a series of corrective actions that the full scope of the defect was identified and addressed.

Six patients. Over a period of approximately two years. The information that the machine was dangerous existed after the first incident. The institutional system — the reporting channels, the manufacturer's investigation process, the regulatory response — did not act on that information with the urgency the situation required, because the system trusted its own engineering. The machine had been tested. The software had been verified. The Therac-20 had operated safely for years. The evidence that something was wrong was treated as anomalous rather than diagnostic — because the alternative, that the machine was fundamentally unsafe, was incompatible with the institution's confidence in its own product.

Captain Smith maintained speed through ice because the ship had never failed. AECL maintained confidence in the Therac-25 because the Therac-20 had never failed. The pattern is the same. The system's track record of success produced an institutional resistance to believing that the system could fail catastrophically. And while the institution resisted, patients continued to be harmed.

> *"Six patients. Over two years. The information that the machine was dangerous existed after the first incident. The institutional system did not act, because trusting its own engineering was easier than accepting that the engineering was killing people."*

The Lesson That Was Not Learned

The Therac-25 accidents led to significant reforms in medical device software regulation. The FDA strengthened its requirements for software verification. AECL redesigned the machine with independent hardware safety interlocks — restoring the keeper that had been removed. The Leveson-Turner paper became required reading in software engineering courses. For a time, the lesson seemed to have penetrated: do not remove independent safety layers. Do not trust software to do what hardware used to guarantee. Do not assume that a system is safe because its predecessor was safe in a different configuration.

And then the lesson was not learned.

In 2018 and 2019, a Boeing 737 MAX — a system that trusted a single sensor to control an automated flight function, with no independent hardware override, in a configuration that the pilots were not fully briefed on — killed 346 people in two crashes. The Therac-25's lesson, published in 1993, taught the world that removing independent safety layers and trusting software to replace them was a recipe for catastrophe. Twenty-five years later, the same architecture killed ten times as many people.

This is the detail that separates this book from a catalog of historical accidents. The Therac-25 is not just a story about a radiation machine in the 1980s. It is a pattern that repeats — in aviation, in autonomous vehicles, in AI-driven medical diagnostics, in military targeting systems — because the economic incentive to remove the independent safety layer (it is expensive, it is slow, it is redundant most of the time) is always stronger than the institutional memory of what happened the last time it was removed.

The keeper is expensive. The keeper is slow. The keeper is unnecessary 99.99% of the time. But the keeper is the difference between a system that fails and recovers and a system that fails and kills. And the Therac-25 proved that in the starkest possible terms, in a treatment room where a patient trusted a machine to heal them and the machine delivered a hundred times the dose because an operator was good at her job and a deadbolt had been replaced by a line of code.

The Pattern Restated

The Titanic trusted its engineering, so the lifeboats were reduced. The Therac-25 trusted its software, so the hardware interlocks were removed. In both cases, the independent safety layer was eliminated because the primary system was judged to be sufficient. In both cases, the judgment was formed by a track record of success under conditions that did not include the specific failure that eventually occurred. In both cases, people died in the gap between what the system was designed for and what actually happened.

The difference between the Titanic and the Therac-25 is the role of the computer. On the Titanic, every failure was a human decision: maintain speed, reduce lifeboats, ignore warnings. On the Therac-25, the lethal failure was a software timing error that no human could perceive, diagnose, or prevent in the moment it occurred. The patient on the table had no warning. The operator at the console had no indication. The error happened inside the machine, at the speed of the processor, in the eight-second window between one keystroke and the next.

The next chapter tells the story of what happens when this dynamic scales from a treatment room to an aircraft carrying 228 people over the Atlantic at 38,000 feet. The Therac-25 removed a hardware interlock. The Airbus A330 removed something more fundamental: the pilot's proficiency at the task the machine had taken over. The interlock was a physical mechanism. The proficiency was a human skill. And when the machine failed and the human was asked to take over, the skill was no longer there.

PART II: THE AUTOMATION PARADOX

The Keeper Who Saved the World

Shortly after midnight on September 26, 1983, Lieutenant Colonel Stanislav Petrov was the duty officer at Serpukhov-15 — the Soviet Union's early-warning command centre, a bunker south of Moscow responsible for monitoring American nuclear missile launches. His job was simple in description and annihilating in consequence: if the system detected an incoming strike, he was to report it immediately to his superiors, who would have minutes to decide whether to launch a retaliatory strike before Soviet weapons were destroyed on the ground.

At 12:14 AM, the system reported a launch. A single American Minuteman intercontinental ballistic missile, inbound. The screen flashed LAUNCH in large red letters. The siren sounded. Within seconds, the system reported a second missile. Then a third. Then a fourth and fifth. Five American nuclear missiles, confirmed by satellite, heading toward the Soviet Union.

Petrov had the training, the authority, and the obligation to report the attack as confirmed. The system was telling him, with high confidence, that the United States had launched a first strike. His commanding officers were waiting. The protocol was clear. The machine had spoken.

He did not report it. He declared it a false alarm.

His reasoning was human in a way that no algorithm could replicate. A genuine American first strike, he judged, would not begin with five missiles. It would begin with hundreds — an overwhelming salvo designed to destroy Soviet retaliatory capability before it could be used. Five missiles made no strategic sense. And the ground-based radar, which should have corroborated the satellite detection, showed nothing. The system was confident. Petrov was not. He trusted his judgment over the machine's certainty, knowing that if he was wrong, the missiles would arrive in approximately twenty minutes and his country would have no warning.

He was right. The satellite system had misidentified sunlight reflecting off high-altitude clouds as missile exhaust plumes. It was a sensor error — the kind of failure that machines produce and humans catch, when the humans are allowed to exercise judgment rather than merely ratify a computation.

"The system was confident. Petrov was not. He trusted his judgment over the machine's certainty, knowing that if he was wrong, the missiles would arrive in twenty minutes and his country would have no warning."

Stanislav Petrov saved the world because he was a keeper who was allowed to keep. He had time — minutes, not seconds. He had context — training in how a real attack would look versus what the machine was showing him. He had authority — the power to override the system based on his own assessment. And he had something no machine possesses: the capacity to recognise that the data, however confidently presented, did not make sense.

The chapters that follow are about what happens when those conditions are removed. When the time is compressed from minutes to seconds. When the context is eroded by automation that deskills the human. When the authority is overridden by a system the operator cannot see inside. When the machine's confidence becomes, structurally, unchallengeable.

In 1983, Petrov had twenty minutes and his own judgment. In the Lavender targeting system deployed forty years later, the human in the loop has twenty seconds and a rubber stamp. The distance between those two numbers is the distance this section measures.

The Last Keeper

The Handoff

On Air France Flight 447, fewer than four minutes
over the Atlantic, the pilots who could not fly the plane the autopilot
had been flying for them, and the most dangerous moment in any
automated system: the handoff from machine to human

*"Children who learn to ride a bicycle with training wheels take
longer to ride without them than children who never had training
wheels at all."*
— observation from motor learning research

The Handoff

On Air France Flight 447, fewer than four minutes over the Atlantic, and the most dangerous moment in any automated system

At 2:10 AM local time on June 1, 2009, the autopilot on Air France Flight 447 disconnected.

The Airbus A330 was at 35,000 feet over the equatorial Atlantic, roughly three and a half hours into a flight from Rio de Janeiro to Paris. Two hundred and twenty-eight people were aboard — 216 passengers and 12 crew. The captain, Marc Dubois, had left the cockpit to rest. In the left seat was first officer David Robert, 37 years old. In the right seat was first officer Pierre-Cédric Bonin, 32 years old. It was dark. The aircraft was passing through the Intertropical Convergence Zone — a belt of thunderstorms that circles the equator — and had entered a region of high-altitude ice crystals.

The ice crystals blocked the aircraft's pitot tubes — small probes on the outside of the fuselage that measure airspeed by detecting the pressure of oncoming air. With the pitot tubes obstructed, the flight computers received inconsistent airspeed data. The autopilot, designed to disengage when its data is unreliable, did what it was designed to do. It handed the aircraft back to the pilots.

This is the handoff. The moment when the automated system, having encountered a condition outside its operating parameters, transfers control to the human. The moment the training wheels come off. The moment the keeper must do the thing the machine has been doing for them, in conditions the keeper has never practiced, with no warning and no preparation.

Three minutes and twenty-eight seconds later, the aircraft hit the surface of the Atlantic Ocean at approximately 11,000 feet per minute. All 228 people aboard were killed.

> *"The autopilot disconnected. Three minutes and twenty-eight seconds later, 228 people were dead. Between those two events was a human being asked to do something the machine had made him forget how to do."*

The Three Minutes

The cockpit voice recorder and flight data recorder were recovered from the ocean floor two years after the crash, at a depth of nearly 13,000 feet. The French Bureau of Investigation and Analysis — the BEA — published a detailed reconstruction of the final minutes. What the recorders revealed is a study in human capability collapsing under precisely the conditions the automation was supposed to prevent.

When the autopilot disconnected, Bonin — the pilot flying — was startled. His first instinct was to pull back on the side stick, pitching the nose of the aircraft upward. This is a natural human reaction to a perceived emergency — pull up, gain altitude, create distance from the ground. It is also, in the situation the aircraft was in, exactly wrong.

An aircraft flies because air flowing over its wings generates lift. If the nose is pitched too high, the angle of the wing relative to the oncoming air — called the angle of attack — becomes too steep. The smooth airflow over the wing breaks apart into turbulence. The wing stops generating lift. This is called an aerodynamic stall, and it is one of the most basic concepts in flight training. The recovery is equally basic: push the nose down to reduce the angle of attack, allow the wing to regain smooth airflow, and the aircraft flies again. Every student pilot learns this in their first weeks of training. Push the nose down. Regain airspeed. Fly.

Bonin pulled the nose up. The aircraft began to stall. The stall warning sounded — a loud, synthetic voice repeating the word "STALL" in the cockpit. Over the next three minutes, the stall warning sounded seventy-five times.

Bonin continued to hold the nose up.

The Paradox

How could a trained airline pilot not recognize a stall? How could a pilot ignore seventy-five stall warnings? How could a pilot maintain an input that was killing the aircraft for three full minutes while the cockpit told him what was happening?

The answer is the automation paradox, and it is the central concept of this chapter.

A modern Airbus autopilot is extraordinarily reliable. It manages the aircraft's speed, altitude, heading, and engine power with a precision that no human pilot can match over sustained periods. On a typical long-haul flight, the autopilot may be engaged for 95% or more of the flight time. The pilots monitor instruments, manage communications, and handle the takeoff and landing. The flying — the actual physical manipulation of the aircraft through the air — is done by the computer.

This arrangement is, on average, safer than having humans fly the entire time. Autopilots do not get tired, do not get distracted, and do not make the small control errors that accumulate during hours of manual flying. The statistical case for automation in aviation is overwhelming. Commercial aviation is the safest form of transportation in human history, and automation is a primary reason.

But here is what the statistics conceal. Every hour the autopilot flies is an hour the pilot does not fly. Every hour the pilot does not fly is an hour of proficiency that erodes. The skills required to hand-fly an aircraft — to sense the aircraft's state through the controls, to recognize subtle deviations, to respond instinctively to unusual attitudes — are *perishable* skills. They degrade without practice in the same way that a musician's technique degrades without rehearsal. The hands forget. The instincts dull. The connection between what the pilot feels and what the aircraft is doing frays.

The automation paradox states: **the more reliable the automated system, the less practiced the human operator becomes at the task, and the worse the human performs when the system fails and they must take over**. The paradox is that the very reliability of the automation — the thing that makes it valuable — is the mechanism that degrades the human capability that is needed when the automation is gone.

Bonin had approximately 2,900 flight hours. He had been trained in stall recovery. He had passed his checkrides. On paper, he was qualified. But the vast majority of his flying hours had been spent monitoring an autopilot, not hand-flying an aircraft at the limits of its flight envelope, at night, over the ocean, with unreliable instruments. The training was in his file. The proficiency was not in his hands.

The Confusion

The cockpit voice recorder reveals a scene of escalating confusion. Robert, in the left seat, did not initially realize that Bonin was holding the nose up. On the Airbus, the two side sticks operate independently — unlike traditional yokes, which are mechanically linked so that one pilot can see and feel what the other is doing. When Bonin pulled back, Robert's stick did not move. Robert had no tactile indication that the other pilot was commanding a nose-up input.

This design choice matters. In an aircraft with mechanically linked controls, one pilot's input is immediately visible and physically felt by the other pilot. The connection is a form of communication — a continuous, wordless awareness of what the other person is doing. The Airbus side sticks removed this connection in favor of a design that was lighter, more compact, and compatible with the fly-by-wire system that the autopilot required. The tradeoff was that the two pilots could give contradictory inputs simultaneously without either one knowing.

For three minutes, that is exactly what happened. Bonin pulled the nose up. Robert, who eventually realized the aircraft was descending, pushed the nose down. The computer algebraically summed their inputs — the opposing commands partially canceling each other out. The aircraft descended.

Captain Dubois, awakened by the initial upset, returned to the cockpit approximately one minute and thirty seconds after the autopilot disconnected. He found his two first officers struggling with contradictory instruments, contradictory inputs, and a stall warning that kept sounding and stopping — stopping because at very low speeds, the aircraft's computers determined that the airspeed data was unreliable and disabled the stall warning, creating the perverse situation in which the warning sounded when the aircraft was partially stalled and went silent when the stall was complete.

In one of the final exchanges captured by the voice recorder, Bonin said: "But I've had the stick back the whole time." It was the first moment the other pilots understood what he had been doing. Robert replied: "No, no, no, don't climb."

Four seconds later, the aircraft struck the water.

The Keeper Who Could Not Keep

There were three keepers in the cockpit of Air France 447. Three trained, licensed airline pilots. The keeper was not absent. The keeper was not distracted. The keeper was not negligent. The keeper was *deskilled* — a word that the aviation safety community uses to describe the degradation of manual flying ability that results from prolonged reliance on automation.

The Therac-25 removed a hardware interlock — a physical safety mechanism. Air France 447 revealed the removal of something less visible and more consequential: the pilot's proficiency at the task the machine had taken over. The interlock was a device. The proficiency was a human capability. And the capability was not removed in a single decision, the way the interlock was removed in the Therac-25 redesign. It was eroded gradually, flight by flight, hour by hour, as the autopilot performed the task and the pilot watched.

This is what makes the automation paradox more insidious than a simple design flaw. The Therac-25's missing interlock was a decision that could be traced to specific engineers at a specific company. The degradation of Bonin's flying skills was not a decision anyone made. It was the cumulative consequence of a system that worked as designed. Nobody decided to deskill the pilots. The automation deskilled them by doing its job too well.

The Reliability Trap Restated

Air France 447 is not an isolated case. It is the most vivid expression of a pattern that aviation safety researchers have documented for decades. In 2013, the FAA published a study that found airline pilots were increasingly unable to fly effectively when the autopilot was unavailable — not because of inadequate training, but because the opportunity to practice had been systematically removed by the automation itself.

Researchers in human factors engineering have a name for this dynamic: the **reliability trap**. A system becomes more reliable. Because it is more reliable, the human operator practices less. Because the human practices less, the human's ability degrades. Because the ability degrades, the consequences of the system failing become more severe. The very success of the automation is the mechanism of the human's decline. The trap is that you cannot escape it by making the system *more* reliable — that only tightens the trap. And you cannot escape it by making the system *less* reliable — that defeats the purpose of the automation.

The reliability trap is not limited to aviation. It is present in every domain where automation performs a task that a human must take over when the automation fails: power plant operations, surgical robotics, autonomous vehicle systems, naval warfare systems. Every domain where the machine is better than the human 99% of the time, and the human must be better than the machine in the 1% that matters most. The asymmetry is structural. The machine practices constantly. The human practices never. And the test comes without warning.

The Question the Industry Does Not Ask

The BEA's final report on Air France 447 was published in 2012. It identified multiple contributing factors: the frozen pitot tubes, the crew's failure to recognize and respond to the stall, the design of the independent side sticks, and the stall warning logic that went silent at full stall. The report led to changes in pilot training, including mandatory high-altitude stall recovery exercises and increased emphasis on manual flying proficiency.

These were important and necessary reforms. They were also insufficient, because they addressed the symptoms without confronting the structural question: if the autopilot is reliable enough that pilots rarely fly manually, and manual flying skills degrade without practice, and the autopilot is designed to disconnect at precisely the moments when conditions are most difficult and manual skills are most needed — then the system contains a contradiction that no amount of training can resolve.

The automation is reliable 99.99% of the time. The handoff occurs during the 0.01% when conditions are worst. The pilot must perform a task they rarely practice, in conditions they have never experienced, with no transition and no warning. The system is designed so that the human is needed most precisely when the human is least prepared.

> *"The system is designed so that the human is needed most precisely when the human is least prepared. The automation paradox is not a bug. It is a structural property of every system that automates a skill."*

The Pattern Deepens

The Titanic's officers had hours to process warnings and chose to trust the ship. The Therac-25's operators had seconds to respond to an error message they could not interpret. Air France 447's pilots had fewer than four minutes to recover an aircraft from a condition that the automation had, in a meaningful sense, trained them not to recognize.

There is a counterexample that makes the mechanism precise. On July 20, 1969, during the lunar descent of Apollo 11, the guidance computer became overwhelmed with data and threw a series of memory alarms — codes 1202 and 1201 — while steering the lunar module toward a boulder-strewn crater. Neil Armstrong took manual control, flew the craft laterally over the hazard, and landed with roughly fifteen seconds of fuel remaining. He could do this because he had not been deskilled by the system he was overriding. He was a test pilot who had practiced manual control thousands of times. When the machine failed, the keeper's hands remembered what to do.

The pilots of Air France 447 had not practiced manual flight at altitude in years. When their machine disconnected, their hands did not remember. The difference between Armstrong and Bonin is not courage. It is practice. And practice is the first thing that automation erodes.

The pattern is accelerating. The time available for human judgment is shrinking. The demands on that judgment are increasing. And the skill required to exercise it is degrading, because the automation that was supposed to assist the human is instead replacing the human's ability to perform without assistance.

In the next chapter, the pattern reaches its most extreme expression yet. The Boeing 737 MAX did not merely fail to help the pilot. It actively fought the pilot. The keeper was in the cockpit, hands on the controls, pulling with everything he had. The machine overrode him. And the machine was acting on data from a single sensor that was wrong.

The Therac-25 removed the keeper and the patient died. Air France 447 degraded the keeper and 228 people died. The 737 MAX overrode the keeper. The arc of these three chapters is the arc of the entire book: from removal to degradation to override. The human is first absent, then unprepared, then powerless. And the consequences escalate with each step.

The Last Keeper

The Override

On the Boeing 737 MAX, a single faulty sensor, an automated system the pilots did not know existed, and 346 people who died because the machine decided it knew better than the humans holding the controls

*"The fault was not that the system overrode the pilot.
The fault was that no one told the pilot the system could."*

The Override

On the Boeing 737 MAX, a single faulty sensor, and 346 people who died because the machine decided it knew better than the pilot

In 2011, Boeing had a problem. Its rival, Airbus, had launched the A320neo — an updated version of the bestselling A320 narrow-body jet, fitted with new, more fuel-efficient engines. The A320neo offered airlines approximately 15% better fuel economy, which over the twenty-to-thirty-year life of a commercial aircraft represents hundreds of millions of dollars. Airlines were placing orders. American Airlines, a longtime Boeing customer, signaled it was considering switching to Airbus.

Boeing faced a choice. It could design an entirely new aircraft — a clean-sheet design with modern aerodynamics and systems — which would take approximately a decade and cost tens of billions of dollars. Or it could do what Airbus had done: take an existing, proven airframe and fit it with new engines. The 737, Boeing's equivalent to the A320, had been in production since 1967. It was the bestselling commercial aircraft in history. Boeing knew it. Airlines knew it. Pilots knew it. Regulators knew it. Fitting new engines to the 737 would be faster, cheaper, and — crucially — would allow Boeing to certify the new aircraft as a variant of the existing 737 rather than as a new type. This meant that airlines could operate the new aircraft without expensive pilot retraining. A pilot certified on the 737 could step into the MAX and fly it the same day.

Boeing chose the second option. The aircraft that resulted was the 737 MAX.

The Physics Problem

The new engines were larger and heavier than the ones they replaced. The 737's airframe sits relatively low to the ground — a design feature that dates to the 1960s, when the aircraft was designed to operate from poorly equipped airports where passengers boarded via stairs rather than jet bridges. The low ground clearance meant the larger engines could not be mounted in the same position as the old ones. They would scrape the runway.

Boeing's engineers moved the engines forward and upward on the wing. This solved the ground clearance problem but introduced an aerodynamic one: in certain flight conditions, particularly at high angles of attack and low speeds, the repositioned engines created additional lift ahead of the aircraft's center of gravity. This had the effect of pitching the nose upward — the opposite of what the pilot intended when pulling back on the yoke. If uncorrected, this nose-up tendency could lead to a stall.

The aerodynamic behavior was subtle. In most flight conditions, the 737 MAX handled identically to the earlier 737 NG. The nose-up tendency appeared only in a narrow band of the flight envelope — high angle of attack, low speed, specific flap configurations. But it was there, and it was a problem, because if the MAX handled differently from the NG in any condition, it might require a separate type rating. That would mean pilot retraining. That would mean airlines would need to invest in new training programs, new simulators, new certification processes. That would eliminate the MAX's key competitive advantage: operational interchangeability with the existing 737 fleet.

"If the MAX handled differently from the NG in any condition, it would need a separate type rating. That would eliminate the aircraft's key competitive advantage. The solution was software."

The Software Solution

Boeing's solution was a software system called the Maneuvering Characteristics Augmentation System — MCAS. The system was designed to monitor the aircraft's angle of attack and, when it detected the nose-up tendency caused by the repositioned engines, automatically push the nose down by adjusting the horizontal stabilizer — the large movable surface at the tail of the aircraft that controls pitch.

In principle, MCAS was a reasonable engineering solution. Software augmentation of flight characteristics is common in modern aircraft. The Airbus fly-by-wire system uses software extensively to shape the aircraft's handling characteristics. The concept was not new.

But the implementation had three features that, in combination, made MCAS lethal.

First, MCAS relied on a single angle-of-attack sensor. The 737 MAX has two angle-of-attack vanes — one on each side of the fuselage. MCAS used only one of them. If that sensor was faulty, MCAS would receive incorrect data and act on it. There was no cross-check, no comparison between the two sensors, no redundancy. A single point of failure controlled a system that could override the pilot's inputs.

Second, MCAS could activate repeatedly. In its original design specification, MCAS was intended to provide a single, limited nose-down correction. During development, the system's authority was increased — the amount of stabilizer movement it could command grew larger, and MCAS was allowed to activate repeatedly, each time adding more nose-down input. If the faulty sensor continued to report a high angle of attack, MCAS would continue to push the nose down, each cycle adding to the cumulative deflection, until the stabilizer reached its mechanical limit. The cumulative force was strong enough to overpower the pilots' ability to pull back on the yoke.

Third, the pilots were not told about MCAS. Boeing did not include MCAS in the 737 MAX's pilot training materials. The system was not described in the aircraft's flight manual. Pilots transitioning from the 737 NG to the 737 MAX were not briefed on MCAS, were not trained to recognize its activation, and were not taught how to disable it. The system existed, it could override their control of the aircraft, and they did not know it was there.

Lion Air Flight 610

On October 29, 2018, Lion Air Flight 610 departed Jakarta, Indonesia, bound for Pangkal Pinang. The aircraft was a 737 MAX 8, delivered to Lion Air just two months earlier. There were 189 people aboard.

Twelve minutes after takeoff, the aircraft crashed into the Java Sea. There were no survivors.

The flight data recorder revealed the following sequence. Shortly after takeoff, the left angle-of-attack sensor began reporting erroneous data — indicating a high angle of attack when the aircraft was flying normally. MCAS, reading this data, activated and pushed the nose down. The captain, feeling the unexpected nose-down movement, pulled back on the yoke to correct it. He succeeded temporarily. Then MCAS activated again. The captain pulled back again. MCAS pushed down again.

This cycle repeated more than twenty times in the eleven minutes between the first MCAS activation and the crash. Each time, MCAS added more nose-down stabilizer. Each time, the captain fought to bring the nose back up. The force required to hold the nose up increased with each cycle as the stabilizer accumulated more deflection. The captain was in a tug-of-war with a machine that had more authority than he did, that could not be reasoned with, that could not be told the sensor was wrong, and that he did not know the name of.

The first officer searched the quick reference handbook for the source of the problem. He could not find it, because MCAS was not in the handbook.

The aircraft hit the water at approximately 450 knots in a near-vertical dive. The impact was so severe that the remains of the passengers could only be identified through DNA analysis.

Ethiopian Airlines Flight 302

On March 10, 2019 — less than five months later — Ethiopian Airlines Flight 302 departed Addis Ababa bound for Nairobi. The aircraft was a 737 MAX 8. There were 157 people aboard, from 35 countries.

Six minutes after takeoff, the aircraft crashed into terrain. There were no survivors.

The flight data recorder revealed the same pattern. A faulty angle-of-attack sensor. MCAS activation. The captain fighting the nose-down commands. MCAS overriding him. The stabilizer accumulating deflection beyond what the pilots could counteract.

Captain Yared Getachew was 29 years old with over 8,000 flight hours. Unlike the Lion Air crew, Getachew and his first officer had the benefit of the worldwide attention that followed the Lion Air crash. Boeing had issued a bulletin after Lion Air reminding pilots of the existing runaway stabilizer procedure. Getachew and his first officer followed that procedure: they flipped the cutout switches that disconnected the electric stabilizer trim, which should have disabled MCAS.

But with the electric trim disabled, the pilots had to move the stabilizer manually using a hand crank in the center pedestal. The problem was that MCAS had already driven the stabilizer to a significant nose-down position. At the aircraft's speed, the aerodynamic forces on the stabilizer were so great that the manual crank could not overcome them. The pilots could not physically turn the wheel. They were strong enough. The physics were not.

In a last, desperate attempt, the crew re-engaged the electric trim, hoping to use its motor to drive the stabilizer back. MCAS immediately activated again and pushed the nose further down. The aircraft entered an unrecoverable dive.

The keepers did everything right. They identified the problem. They followed the published procedure. They flipped the correct switches. And they died because the machine had already moved the control surface beyond the point where human strength could recover it, and nobody at Boeing had tested whether the published procedure was physically executable after MCAS had been active for more than a few seconds.

The Architecture of the Failure

The US House Transportation Committee's investigation, published in 2020 after an 18-month inquiry, documented a failure that extended from the engineering floor to the boardroom to the regulatory agency that was supposed to provide independent oversight.

Boeing had designed MCAS to solve a business problem as much as an engineering one. The aerodynamic issue was real, but the decision to solve it with a software patch rather than a redesign was driven by schedule pressure and the need to maintain the 737 MAX's certification status as a variant of the 737 NG. Every design choice — single sensor, no pilot notification, expanded authority — was made in the direction of minimizing the regulatory and training impact of the system.

The FAA, the regulatory body responsible for certifying the aircraft as safe to fly, had delegated significant portions of the certification process to Boeing itself. This was not unusual — the FAA had used delegated certification for years as a way of managing its workload. But the effect was that Boeing employees were simultaneously designing the system and approving the design. The engineers who determined that MCAS was safe were, in many cases, the same engineers who had designed it. The independent safety layer — an external regulator with the authority and expertise to challenge Boeing's assessment — had been thinned to the point of transparency.

Boeing's own internal communications, revealed during the Congressional investigation, showed that some employees recognized the risks. One test pilot wrote in an internal message that MCAS was "running rampant" in a simulator session. Another employee, discussing the decision not to include MCAS in pilot training, expressed concern about the system's potential behavior. These concerns did not change the certification outcome. The schedule held. The aircraft was delivered.

This is the institutional pattern from the Therac-25, scaled upward. AECL could not reproduce the error and declared the machine safe. Boeing's analysis concluded MCAS was safe. In both cases, the manufacturer's confidence in its own engineering overrode the evidence — and in both cases, an independent safety layer that should have caught the error had been weakened: hardware interlocks removed in the Therac-25, regulatory independence compromised in the 737 MAX. The pattern does not vary. The institution trusts itself. The backup is eroded. People die in the gap.

Removal, Degradation, Override

The three chapters that form Part II of this book trace an arc that defines the relationship between humans and automated systems.

In Chapter 3, the keeper was **removed**. The Therac-25's hardware interlocks — independent, physical, reliable — were taken away because the software was judged to be sufficient. The human operator was present but had no means of detecting or preventing the error. The safety layer between the bug and the patient was gone.

In Chapter 4, the keeper was **degraded**. Air France 447's pilots were in the cockpit, hands on the controls, trained and licensed. But the automation that was supposed to assist them had, over thousands of hours, eroded the manual flying skills they needed when the automation departed. The keeper was present but had been rendered unable to keep.

In this chapter, the keeper was **overridden**. The 737 MAX pilots were present, aware that something was wrong, and actively fighting to save the aircraft. The machine fought back. The machine had more authority than the humans. The machine was acting on a single faulty data point. And the machine won.

Removal. Degradation. Override. This is the trajectory. Each step is more insidious than the last, because each step leaves the human with the *appearance* of control while reducing the *substance* of control. The Therac-25 operator appeared to control the treatment but could not detect the error. The Air France pilots appeared to be flying but had forgotten how. The 737 MAX pilots were fighting for their lives and losing to a system they did not know existed.

> *"Removal. Degradation. Override. Each step leaves the human with the appearance of control while reducing the substance of control. The human is first absent, then unprepared, then powerless."*

Beyond the Cockpit

The next three chapters leave the world of physical systems — machines, aircraft, medical devices — and enter the world of algorithmic systems. The failures in Part II killed people through physics: radiation beams, aerodynamic stalls, nose-down forces that exceeded human strength. The failures in Part III will harm people through decisions: who goes to prison, who receives medical care, who is shown content that radicalizes them toward violence.

The mechanism changes. The pattern does not. In every case, an automated system replaces or overrides human judgment. In every case, the system operates on data that is incomplete, biased, or wrong. The humans who are harmed have no way to see inside the system, challenge its reasoning, or appeal its decision. And the keeper — the human who should exercise independent judgment at the point of consequence — has been removed, degraded, or overridden.

The cockpit is a constrained environment. The angle of attack is a physical measurement. The stall is a physical phenomenon. The crash is a physical event. The next chapters enter a domain where the inputs are social, the outputs are human lives, and the errors are not crashes but injustices — quieter, slower, distributed across millions of individual decisions, and for exactly those reasons, harder to see and harder to stop.

The Last Keeper

Chapter 6 of 12

The Orbit

On the Ariane 5 rocket destroyed by a software conversion error,
the Mars orbiter lost to a unit mismatch, and the question of
what happens when the systems we build exceed our capacity
to verify them

*"The code worked exactly as written.
It was written for a different rocket."*

The Orbit

On the systems that exceeded our capacity to verify them

The previous chapters described systems where the keeper was removed and the failure was, in retrospect, legible. The Titanic's officers maintained speed through ice — a decision traceable to a culture of trust in engineering. The Therac-25's race condition was a specific flaw in specific code, found and fixed once investigators knew where to look. Air France 447's pilots could not fly the plane — a skill gap produced by years of reliable automation. The 737 MAX fought its pilots because of a single-sensor dependency that Boeing chose not to disclose.

In each case, the failure had a name. A cause. A line of code, a design decision, a regulatory gap that could be pointed to and said: *there*. The keeper was removed, and the removal was identifiable. The lesson, however painfully learned, was learnable.

This chapter is about what happens when the failure stops being legible. When the systems we build grow past the point where human beings can verify that they will work correctly under all the conditions they will encounter. When the keeper may be present, empowered, skilled, and willing — but the system is too complex for the keeper to understand. This is a different kind of threat than a removed keeper. It is a keeper who *cannot* keep, because the thing being kept has crossed a threshold beyond which human comprehension does not reach.

Thirty-Seven Seconds

On June 4, 1996, the European Space Agency launched the Ariane 5 rocket on its maiden flight from the Guiana Space Centre in Kourou, French Guiana. The rocket carried four Cluster satellites — a scientific mission to study the interaction between the solar wind and the Earth's magnetosphere. The total cost of the payload and launch vehicle was approximately $500 million.

Thirty-seven seconds after liftoff, the rocket veered off course, broke apart under aerodynamic stress, and self-destructed. The payload was lost.

The Cluster mission had been ten years in planning. Hundreds of scientists across fourteen European countries had designed the instruments, calibrated the sensors, and waited for the data that would advance humanity's understanding of space weather. The principal investigators had spent careers building toward this launch. They watched from the control room in Toulouse as a decade of work disintegrated in thirty-seven seconds. The mission was eventually reflown — Cluster II launched successfully in 2000 — but the four years of delay, the loss of irreplaceable instruments, and the human cost of watching your life's work explode are not captured in the technical reports.

The subsequent investigation, led by an inquiry board chaired by Jacques-Louis Lions, identified the cause with unusual precision. The failure began in the rocket's inertial reference system — the component that measures the rocket's orientation and velocity. The software in the inertial reference system attempted to convert a 64-bit floating-point number, representing the rocket's horizontal velocity, into a 16-bit signed integer. The number was too large to fit. The conversion produced an overflow error. The error handler, following its   d instructions, shut down the inertial reference system.

The backup inertial reference system — the redundant unit, the safety layer — had already failed, 72 milliseconds earlier, from the identical error. Both units ran the same software. Both encountered the same overflow. The redundancy was an illusion: two copies of the same code, executing the same flaw, failing in the same way, 72 milliseconds apart.

"The backup system failed 72 milliseconds before the primary system, from the identical error. Two copies of the same code. The same flaw. The redundancy was an illusion."

With both inertial reference systems offline, the flight computer received no valid orientation data. It interpreted the error codes from the failed systems as flight data — data indicating that the rocket needed a massive and immediate course correction. The nozzles of the solid rocket boosters swiveled to their maximum deflection. The rocket turned sharply. The aerodynamic forces exceeded the vehicle's structural limits. The rocket broke apart. The self-destruct system activated.

The critical detail: the software that caused the failure was not written for the Ariane 5. It was reused from the Ariane 4, the previous rocket in the program. The Ariane 4 software had worked correctly for years. The conversion that failed on the Ariane 5 had never failed on the Ariane 4, because the Ariane 4's trajectory produced horizontal velocity values that were small enough to fit in a 16-bit integer. The Ariane 5's trajectory was different — steeper, faster, producing larger velocity values that exceeded the integer's range.

The code worked exactly as written. It was written for a different rocket.

The Review That Did Not Catch It

The Ariane 5 software was not unreviewed. It was tested. It was verified. The European Space Agency employed rigorous development and review processes. The reuse of Ariane 4 software was a deliberate decision, made on the reasonable basis that the software had a proven track record.

The Lions report identified that the specific conversion from 64-bit floating-point to 16-bit integer had been analyzed during the Ariane 4 program and determined to be safe — because the Ariane 4's flight parameters guaranteed that the value would never exceed the 16-bit range. When the software was transferred to the Ariane 5, this analysis was not repeated. The assumption that the value would remain within range was inherited along with the code. The assumption was valid for the old rocket. It was not valid for the new one.

The reader who has been following this book will recognize the architecture. The Therac-25 reused software from the Therac-20. The software had a flaw that was harmless on the old machine because the hardware interlocks prevented it from manifesting. On the new machine, the interlocks were removed, and the flaw became lethal. The Ariane 5 reused software from the Ariane 4. The software had a limitation that was harmless on the old rocket because the flight profile kept the values within range. On the new rocket, the flight profile was different, and the limitation became catastrophic.

In both cases, the system was reviewed. In both cases, the reviews did not catch the flaw. Not because the reviewers were incompetent, but because verifying that inherited code is safe in a new context requires understanding every assumption embedded in the inherited code and testing every assumption against the new context. In simple systems, this is feasible. In complex systems, the number of embedded assumptions exceeds the capacity of any review team to enumerate, let alone test.

The Unit

Three years after the Ariane 5 disaster, NASA lost the Mars Climate Orbiter — a $328 million spacecraft designed to study the Martian atmosphere and serve as a communications relay for future missions. On September 23, 1999, the spacecraft approached Mars for its orbital insertion maneuver and was never heard from again. It had entered the atmosphere at an altitude of approximately 57 kilometers instead of the planned 226 kilometers — too low to survive.

The investigation revealed that one engineering team, at Lockheed Martin, had been providing thruster performance data in imperial units — pound-force seconds. The navigation team at NASA's Jet Propulsion Laboratory expected the data in metric units — newton-seconds. The difference is a factor of approximately 4.45. Over the nine months of the spacecraft's cruise to Mars, the accumulated error from this mismatch shifted the trajectory enough to destroy the mission.

A unit conversion error. Not a novel algorithm. Not an unprecedented engineering challenge. A conversion between pounds and newtons — the kind of calculation a physics student learns in their first semester. Three hundred and twenty-eight million dollars lost because one team said pounds and another team heard newtons, and no process in between caught the discrepancy.

The Mars Climate Orbiter failure is, in isolation, almost comical — the kind of error that seems impossible at the level of sophistication involved. But it illustrates a principle that the Ariane 5 failure also illustrates, and that this chapter exists to name: **the failures in complex systems are rarely complex**. They are simple errors — a data type conversion, a unit mismatch, an inherited assumption — embedded in systems so complex that the simple error is invisible.

The Complexity Threshold

The Ariane 5's flight software contained approximately several hundred thousand lines of code. This was considered complex at the time. It was reviewable — with difficulty, with significant effort, and with the gaps that the Lions report identified — but reviewable. A team of engineers could, in principle, read every line, trace every data flow, and test every conversion. The fact that they did not catch the fatal conversion was a failure of process, not of possibility.

Modern software systems operate at a different scale. The Linux kernel — the operating system that runs the majority of the world's servers, including those hosting cloud services, financial systems, and critical infrastructure — contains over 30 million lines of code. A modern automobile contains approximately 100 million lines. The Google codebase is estimated at over 2 billion lines. No individual can read, understand, or verify these systems in their entirety. No team can. The systems are maintained by organizations that understand their respective components, with no single person or group holding a complete understanding of the whole.

These are systems whose behavior is, in the most literal sense, beyond human comprehension in totality. We can understand the parts. We can test the interfaces between parts. We can employ automated testing tools that exercise millions of code paths. But the space of possible interactions between components in a system of 100 million lines of code is combinatorially vast — larger than the number of atoms in the universe. Testing covers a fraction. Review covers a fraction. The rest is trust.

The Black Box

And then there are neural networks.

A large language model — the kind of AI system that powers modern chatbots, code generators, and an increasing number of decision-support tools in medicine, law, finance, and government — contains billions of parameters. These parameters are numerical values that were not written by engineers. They were learned during training — adjusted through millions of iterations of a mathematical optimization process that modifies the values to improve the system's performance on its training data.

The resulting system works. It produces outputs that are, much of the time, remarkably capable. But no human designed those parameters. No human can explain, for a given output, exactly which parameters contributed to the result and why. The system is not opaque because its creators chose to hide the logic. It is opaque because there is no logic in the human sense — no sequence of if-then rules, no decision tree, no flowchart. There are billions of numbers, arranged in layers, that collectively produce behavior. The behavior works. The mechanism is beyond human interpretation.

The Ariane 5's fatal code was readable. An engineer could open the file, find the conversion, and see the flaw. The flaw was not found in time, but it was findable. A neural network's equivalent of the Ariane 5 flaw — the embedded assumption that will produce catastrophic failure under conditions the system was not trained on — is not readable. It is distributed across billions of parameters in ways that no current technique can fully trace. The keeper who wants to verify the system cannot verify it. Not because the keeper lacks access, or skill, or time. Because the system's complexity has crossed a threshold beyond which human verification is not difficult but definitionally impossible.

The Implication

This chapter exists to establish a fact that the remaining chapters of this book must confront. The systems now being deployed in the most consequential domains of human life — medical diagnosis, criminal sentencing, military targeting, financial markets, critical infrastructure — are systems that no human being can fully verify.

This is not a temporary condition that better tools will resolve. It is a structural property of the systems themselves. A neural network with 175 billion parameters is not a system that is difficult to interpret. It is a system that produces behavior from a mechanism that has no human-readable explanation. We can probe it. We can test it. We can observe its outputs and characterize its behavior statistically. But we cannot look inside it the way Jacques-Louis Lions looked inside the Ariane 5's code and said: here is the conversion, here is the overflow, here is the fault.

The Therac-25's flaw was a race condition. It was found, identified, and fixed. The 737 MAX's flaw was a single-sensor dependency. It was found, identified, and fixed. The Ariane 5's flaw was a data type conversion. It was found, identified, and fixed. In each case, the flaw was legible — expressible in language, traceable to a specific line of code or a specific design decision. The fix was possible because the flaw was comprehensible.

What do we do when the flaw is not comprehensible? When the system produces a wrong output — a misdiagnosis, a wrongful recommendation, a catastrophic financial trade — and no one can explain why? When the keeper looks inside the system and finds not a readable error but an impenetrable matrix of numbers that produced the wrong answer for reasons that cannot be expressed in language?

The chapters that follow will document what happens when systems of this complexity are deployed in domains where the stakes are measured in lives, liberty, and justice. Algorithms that sentence defendants. Healthcare systems that allocate care. Recommendation engines that shape what billions of people believe. Targeting systems that select who dies. Surveillance infrastructure that monitors entire populations. In each case, the keeper will face the same question this chapter has raised: how do you keep a system you cannot understand?

> *"How do you keep a system you cannot understand? The Ariane 5's flaw was findable. A neural network's equivalent flaw is distributed across billions of parameters. The system has crossed a threshold beyond which human verification is not difficult but impossible."*

What Has Been Established

Take stock of what this book has documented so far.

A ship whose officers trusted the engineering more than the warnings, and 1,517 people died in the Atlantic. A radiation machine whose hardware safety layer was removed because the software was believed to be sufficient, and patients received a hundred times the intended dose. An aircraft whose pilots could not fly because the autopilot had made flying unnecessary, and 228 people fell from the sky. A second aircraft whose automated system fought its pilots to the death because it trusted a single faulty sensor. And now: a rocket and a spacecraft destroyed not by malice or negligence but by the simple, inevitable consequence of building systems too complex for their creators to fully verify.

The pattern has escalated. The Titanic was a failure of human judgment — the keeper was present but calibrated to a system that had never failed. The Therac-25 was a failure of software design — the keeper was removed and the flaw was invisible. Air France 447 was a failure of human skill — the keeper was deskilled by the very automation that was supposed to help. The 737 MAX was a failure of institutional honesty — the keeper was denied the information needed to keep. The Ariane 5 and Mars Orbiter are failures of complexity itself — the keeper cannot verify what the keeper cannot comprehend.

And the systems that follow in the remaining chapters of this book — the algorithms that judge, the engines that amplify, the machines that target, the infrastructure that watches — are systems built on exactly this kind of unverifiable complexity. The keeper's challenge is about to get harder.

The light at Smalls still shines. The mechanism is more reliable than Thomas Howell could have maintained it. The question is not whether the mechanism is better on average. The question has never been about the average. The question is about the night the mechanism fails. And whether, when that night comes, anyone is watching — and whether the watcher can understand what has gone wrong.

The Last Keeper

The Sentencing Machine

On the COMPAS algorithm, the courtrooms that trusted software
to predict who would commit a crime, and the defendant who
could not cross-examine the code that determined his future

*"The question is not whether the algorithm is biased.
The question is whether you can look it in the eye and ask why."*

The Sentencing Machine

On the COMPAS algorithm, the courtrooms that trusted software to predict who would commit a crime, and the defendant who could not cross-examine the code that determined his future

The previous three chapters told stories about machines that killed people. The beam fired. The aircraft crashed. The stabilizer drove the nose into the ground. In each case, the moment of failure was physical, immediate, and unmistakable. The patient felt the jolt. The cockpit voice recorder captured the final seconds. The wreckage was recovered from the ocean floor.

This chapter tells a different kind of story. Nobody dies. Nobody is physically injured. The machine does not malfunction. The machine works exactly as designed, produces the output it was built to produce, and that output is used by human institutions to make decisions that alter the course of human lives — who goes to prison, for how long, under what conditions. The harm is not a beam or a crash. The harm is a number on a screen, presented to a judge, treated as evidence, and used to justify keeping a human being in a cage.

The shift from Part II to Part III is a shift in what failure means. In Part II, the machine failed and people died. In Part III, the machine succeeds — does exactly what it was designed to do — and people are harmed anyway, because the design itself encodes a harm that the designers did not intend and the operators cannot see.

The Promise

The idea behind algorithmic risk assessment in criminal justice is, on its surface, reasonable. Human judges are inconsistent. Studies have shown that sentencing outcomes vary with the judge's mood, the time of day, the order in which cases are heard, and — most troublingly — the defendant's race. A landmark 2018 study of federal sentencing data found that Black men received sentences roughly 19% longer than white men convicted of similar offenses, after controlling for criminal history and offense severity. Judicial discretion, applied inconsistently, produces injustice.

An algorithm, the argument goes, can do better. Feed it data — criminal history, demographic information, behavioral indicators — and it will produce a risk score: a number that predicts, with some statistical accuracy, the likelihood that a defendant will reoffend. The score is consistent. It does not depend on whether the judge had lunch. It does not depend on the defendant's appearance or demeanor. It produces the same output for the same input, every time. It is, in a word, objective.

This was the promise of COMPAS — the Correctional Offender Management Profiling for Alternative Sanctions system — developed by a company called Northpointe (later Equivant). COMPAS uses a 137-question assessment to generate risk scores for criminal defendants. The scores are presented to judges on a scale from 1 to 10, where higher numbers indicate higher predicted risk of recidivism. Judges in courtrooms across the United States use these scores to inform decisions about bail, sentencing, and parole.

The promise was objectivity. The promise was consistency. The promise was that the machine would be fairer than the human. And for a time, the promise was believed.

> *"The promise was that the machine would be fairer than the human. An algorithm cannot be racist. An algorithm does not have a mood. An algorithm does not skip lunch."*

The Investigation

In 2016, a team of journalists at ProPublica published an investigation that would become one of the most cited pieces of data journalism in history. Julia Angwin, Jeff Larson, Surya Mattu, and Lauren Kirchner obtained COMPAS risk scores for more than 7,000 defendants arrested in Broward County, Florida, between 2013 and 2014. They then tracked which of those defendants actually went on to commit new crimes over the following two years, and compared the algorithm's predictions against reality.

The findings were stark.

COMPAS was roughly as accurate as a coin flip for predicting violent recidivism — about 61% overall accuracy, which subsequent research showed was comparable to the prediction accuracy of untrained volunteers given nothing more than the defendant's age and number of prior convictions. The algorithm's advantage over human judgment was, at best, marginal.

But the errors were not distributed equally.

Black defendants who did not go on to reoffend were nearly twice as likely to be incorrectly classified as high risk compared to white defendants who did not reoffend. White defendants who did go on to reoffend were nearly twice as likely to be incorrectly classified as low risk compared to Black defendants who did reoffend. The algorithm was wrong in both directions — but it was wrong in a way that systematically overpredicted danger for Black defendants and systematically underpredicted danger for white defendants.

The algorithm did not use race as an input variable. Northpointe was emphatic on this point. Race was not in the questionnaire. Race was not a factor in the score. The algorithm was, in the company's framing, race-neutral.

This is where the chapter arrives at the concept it exists to explain.

The Proxy

In statistics, a **proxy variable** is a variable that is not itself the thing you are measuring but that correlates closely enough with it to serve as a stand-in. Your zip code is not your race. But in the United States, where decades of redlining, discriminatory lending, and housing segregation have sorted people into neighborhoods along racial lines, your zip code is a remarkably effective proxy for your race. Your employment history is not your race. But in a country where the unemployment rate for Black Americans has been roughly double the rate for white Americans for every year since the Bureau of Labor Statistics began tracking it, your employment history carries racial information. Your number of prior arrests is not your race. But in a country where Black Americans are arrested at 2.6 times the rate of white Americans — a disparity driven in significant part by differential policing of Black neighborhoods — your arrest record carries racial information.

COMPAS did not use race. It used proxies for race. And the proxies worked, in the sense that they produced scores that correlated with race in precisely the way that direct use of race would have. The algorithm achieved the outcome of racial discrimination without the mechanism of racial discrimination. The bias was invisible because it was structural, embedded in the correlations between variables rather than in any single variable.

This is Pattern 3 — the Invisible Bias — in its purest form. The bias is not in the algorithm. The bias is in the data. The data reflects the world as it is — a world shaped by centuries of racial discrimination in housing, employment, education, and policing. The algorithm learns the patterns in the data. The patterns in the data include the effects of that discrimination. The algorithm reproduces the discrimination, not because it was instructed to, but because the discrimination is the pattern.

Imagine teaching a child to sort photographs into two categories: "likely trustworthy" and "likely untrustworthy." You never mention race. But all the photographs labeled "untrustworthy" in your training set happen to be of people from the same neighborhood, which happens to be predominantly Black, which happens to be over-policed, which happens to produce more arrest records, which happens to be the training data the algorithm learned from. The child — and the algorithm — learns the correlation. Not because you taught the correlation. Because the correlation was in the world you showed them.

The Defendant

In 2013, Eric Loomis was arrested in La Crosse, Wisconsin, for driving a car that had been used in a shooting. He was not accused of the shooting itself. He pleaded guilty to eluding a police officer and operating a vehicle without the owner's consent. At sentencing, the judge cited Loomis's COMPAS risk score — which rated him as high risk — as a factor in imposing a sentence of six years of initial confinement.

Loomis challenged the sentence. His argument was straightforward: the COMPAS algorithm is proprietary. Its methodology, its weighting of variables, and its validation data are trade secrets owned by Northpointe. Loomis could not see how his score was calculated. He could not challenge the algorithm's assumptions. He could not present evidence that the algorithm was flawed. He could not, in the most fundamental sense, confront his accuser — because his accuser was a black box owned by a private company that refused to disclose how the box worked.

The case reached the Wisconsin Supreme Court. In 2016, the court ruled that the use of COMPAS in sentencing did not violate Loomis's due process rights. The court acknowledged that the proprietary nature of the algorithm was a concern. The court acknowledged that COMPAS scores correlate with race. The court acknowledged that the defendant could not fully understand or challenge the score. And the court ruled that the score could be used anyway, as long as it was not the sole basis for the sentence.

Read that again. The court said: we know the algorithm is opaque. We know it correlates with race. We know the defendant cannot challenge it. We will allow it anyway.

The US Supreme Court declined to hear Loomis's appeal.

The Impossibility

In the aftermath of the ProPublica investigation, a remarkable academic debate unfolded — one that revealed something troubling not just about COMPAS but about the mathematical foundations of algorithmic fairness itself.

Northpointe responded to ProPublica's findings by arguing that COMPAS was, in fact, fair. Their definition of fairness was different from ProPublica's. Northpointe argued that among defendants who received the same risk score, the actual recidivism rates were similar across racial groups. A Black defendant scored as high risk and a white defendant scored as high risk had approximately the same probability of actually reoffending. The score meant the same thing regardless of race. This is called **predictive parity**.

ProPublica's definition of fairness was different. They argued that among defendants who did not reoffend, Black defendants were far more likely to have been incorrectly classified as high risk. The error rate — the rate at which the algorithm was wrong — was not equal across racial groups. This is called **equal false positive rates**, or **error rate balance**.

Here is the devastating mathematical result: in 2016, researchers proved that in any population where the base rates of the predicted outcome differ between groups — which is to say, in any population where the groups are not identical — it is mathematically impossible to satisfy both definitions of fairness simultaneously. You cannot have predictive parity and equal error rates at the same time, unless the groups have identical base rates of the outcome being predicted.

In the United States, Black Americans are rearrested at higher rates than white Americans. This is a statistical fact. It is also a fact that is itself produced by the same system of differential policing, prosecution, and incarceration that the algorithm is supposed to transcend. The base rates are unequal because the system is unequal. And because the base rates are unequal, it is mathematically impossible to build an algorithm that is fair by both reasonable definitions of fairness.

This is not a software bug. This is not a fixable flaw. This is a mathematical proof that algorithmic fairness, in a society with unequal outcomes across groups, is a contradiction in terms. You must choose which kind of unfairness you prefer. And the choice is not a technical decision. It is a moral one. It is precisely the kind of decision that requires a keeper — a human being who understands the context, weighs the competing values, and takes responsibility for the outcome.

The Missing Keeper

In the courtroom, the keeper is the judge.

A human judge is imperfect. A human judge is inconsistent. A human judge has bad days and blind spots and unconscious biases. All of this is true, and all of it was the justification for introducing algorithmic risk assessment in the first place.

But a human judge can be cross-examined. A human judge's reasoning can be reviewed on appeal. A human judge can be asked to explain why they believed this defendant was dangerous and that defendant was not. A human judge can look at a score and say: I know what the number says, but I have sat across from this person, I have heard their story, I have seen things the questionnaire does not capture, and I believe the number is wrong. A human judge can exercise judgment. That is, in fact, why we call them judges.

COMPAS does not eliminate the judge. The judge still pronounces the sentence. But COMPAS shifts the epistemic ground. When a judge has a number in front of them — a number produced by a system described as scientific, validated on data, endorsed by the corrections department — the burden shifts. The judge who follows the score needs no justification. The judge who overrides the score must explain why. The path of least resistance is to defer to the number. And the number, as we have seen, carries the biases of the world that produced it.

The keeper is still in the room. But the keeper's authority has been hollowed out. The keeper has been given a tool that looks like evidence, feels like objectivity, and carries the weight of institutional endorsement — and overriding that tool requires the judge to assert their own fallible human judgment against the machine's output. This is the degradation pattern from Chapter 4, translated from the cockpit to the courtroom. The pilots of Air France 447 were present but deskilled. The judge in a COMPAS courtroom is present but deauthorized — not formally stripped of power, but structurally discouraged from exercising it.

The Scale

COMPAS is not an outlier. It is one system in an expanding ecosystem of algorithmic decision-making in criminal justice. Risk assessment tools are used in pretrial detention decisions, sentencing, parole hearings, and probation conditions across the majority of US states. Some jurisdictions use tools developed by private companies. Others use tools developed by academic institutions or government agencies. The specific algorithms vary. The structural dynamics do not.

The algorithm takes data that reflects the history of the criminal justice system — including its documented racial disparities — and uses that data to make predictions about the future. The predictions are presented to human decision-makers as objective inputs. The human decision-maker faces the same structural pressure: defer to the number, because overriding it requires justification and deference does not.

The number of people affected is not small. In the United States alone, approximately 1.9 million people are held in prisons, jails, and detention facilities on any given day, and people enter jail more than seven million times each year. Algorithmic risk assessment touches a significant and growing fraction of these cases. Each case is a human being whose liberty is at stake. Each case is a decision that a machine has influenced and that the affected person cannot fully understand, challenge, or appeal.

The Deeper Problem

There is a question beneath the question of algorithmic fairness, and it is the question that matters.

Should we predict crime at all?

The entire premise of risk assessment — algorithmic or otherwise — is that we can identify, before the fact, who is likely to commit a crime and act on that prediction. The prediction is not a finding of guilt. It is an assessment of probability. And acting on a probability — keeping someone in jail, denying bail, imposing a longer sentence — because of what they *might* do is fundamentally different from punishing them for what they *did*.

The philosophical tradition that underlies the Western legal system is built on the principle that punishment follows action. You are punished for what you have done, not for what a statistical model predicts you will do. Risk assessment inverts this. It punishes the prediction. And the prediction is, as we have seen, shaped by the very inequalities it claims to transcend.

This inversion matters, because the legal framework makes it unambiguous. The presumption of innocence is not a technicality. It is the bedrock of Western constitutional law — enshrined in Article 11 of the Universal Declaration of Human Rights, the foundation of the Fifth, Sixth, and Fourteenth Amendments to the United States Constitution. A person cannot be punished for something they have not done. COMPAS inverts this. A defendant who receives a high risk score serves additional time — longer pretrial detention, harsher sentencing, denied parole — based on a statistical prediction about their future behavior. They are being punished not for a crime they committed but for a crime the algorithm predicts they *will* commit, based on characteristics they share with a demographic group. This is statistical profiling dressed as risk assessment. It is the presumption of guilt, computed.

Nobody is challenging this constitutionally, and the reason is the same reason that recurs throughout this book: the people affected lack the resources to litigate it. The populations most harmed by recidivism algorithms are disproportionately Black, disproportionately poor, and disproportionately represented by overworked public defenders who do not have the bandwidth to mount a constitutional challenge to the algorithmic infrastructure of the criminal justice system. The constitutional violation is hiding in plain sight because the victims cannot afford to make it visible.

And this logic does not stop at the courtroom. The same reasoning that says "this person is 60% likely to reoffend, so we keep them locked up longer" is the reasoning that says "this building is 70% likely to contain a combatant, so we strike it." Both treat actual human beings as probabilities. Both invert the presumption of innocence. One operates in a courtroom in Wisconsin. The other operates from a drone over Gaza. The constitutional violation is the same. The altitude is different.

A human judge who admits that a defendant's zip code influenced their sentencing decision would face immediate scrutiny. An algorithm that uses zip code as one of 137 variables produces a number that carries the same influence through a mechanism so complex that the influence is undetectable to the person who receives it. The judge sees a 7 out of 10. The judge does not see the zip code inside the 7.

> *"The judge sees a 7 out of 10. The judge does not see the zip code inside the 7. The algorithm does not hide the bias. It dissolves the bias into a number, which is a more effective way of hiding it."*

The Pattern Crosses the Threshold

In the Therac-25, the operator could not see the race condition. In Air France 447, the pilots could not see each other's inputs. In the 737 MAX, the pilots could not see MCAS. the failure was invisible to the person who needed to prevent it.

COMPAS applies this architecture to human liberty. The judge cannot see inside the algorithm. The defendant cannot see inside the algorithm. The public cannot see inside the algorithm. The algorithm's producer claims trade secret protection. And the courts have ruled that this opacity is permissible.

But there is a difference, and it is the difference that makes Part III of this book more disturbing than Part II. The Therac-25 malfunctioned. The autopilot disconnected. MCAS received bad data. In each case, the system did something it was not supposed to do. COMPAS does exactly what it is supposed to do. It produces a risk score from the data it is given. The problem is not that the system fails. The problem is that the system succeeds — and the success encodes a harm that is built into the structure of the data, the structure of the question, and the structure of the society that produced both.

The keeper is not fighting a malfunction. The keeper is fighting a system that is working as intended, backed by institutional authority, presented as objective, and resistant to challenge. That is a harder fight. And it is the fight that defines the remaining chapters of this book.

But this chapter must be honest about a tension in its own argument. The COMPAS algorithm did not invent racial bias in sentencing. It inherited it. The human judges whose decisions trained the algorithm were themselves biased — and their bias was invisible, unchallenged, and unquantified for decades. Keeping the human judge in the loop does not eliminate the bias. It preserves it in a form that is harder to measure and harder to correct. It was the algorithm — consistent, examinable, producing the same output from the same input — that made the bias *visible* for the first time. ProPublica could measure the racial disparity precisely *because* the algorithm produced measurable outputs. The human judges' individual, opaque reasoning had never been subjected to equivalent scrutiny.

The case for keeping the human in the loop is therefore not a case for the status quo. It is a case for the human who can recognize the bias and correct it — which requires the bias to be exposed. And sometimes it takes an algorithm to expose it. The keeper's value is not neutrality. The keeper is not neutral. The keeper's value is the capacity for recognition and correction that no algorithm possesses on its own.

But there is a harder counterpoint, and intellectual honesty requires naming it. On April 26, 1986, at the Chernobyl Nuclear Power Plant in Ukraine, human operators running a safety test deliberately disabled the automated emergency cooling systems, overrode hardware interlocks, and pushed the reactor into an unstable state that caused the explosion. At Chernobyl, the keepers were not overridden by the machine. The keepers overrode the machine — and the machine's constraints were the only thing that had been preventing catastrophe. The keeper is not inherently safe. Human judgment is only valuable if the human is exercising wisdom. When the keeper acts with arrogance, with recklessness, with the conviction that they know better than the system they are disabling, the automated constraint becomes the last line of defense. This book argues for the keeper. Chernobyl reminds us that the argument is conditional.

From the Courtroom to the Clinic

COMPAS decides who is dangerous. The next chapter tells the story of an algorithm that decides who is sick — and the patients it systematically determined were healthier than they actually were, because the data it was trained on confused spending money on healthcare with needing healthcare. In a country where Black patients have historically received less care, the algorithm learned that Black patients need less care. The algorithm was wrong. The patients suffered. And the bias, once again, was invisible to the humans who trusted the output.

The sentencing machine predicted futures. The diagnostic will misread the present. The mechanism is the same: data that reflects inequality, processed by an algorithm that reproduces inequality, presented to a human who trusts the output because the output looks like science.

The Last Keeper

The Diagnostic

On a nephrologist in Scarborough who looked beneath the categories,
the algorithm that confused cost with need,
and the question of what we measure when we think
we are measuring health

*"The algorithm did not decide that Black patients were healthier.
It decided that patients who cost less to treat were healthier.
In America, those are the same decision."*

The Diagnostic

On a nephrologist in Scarborough who looked beneath the categories, the algorithm that confused cost with need, and the question of what we measure when we think we are measuring health

Scarborough sits on the eastern edge of Toronto. Sixty percent of its residents are immigrants. Seventy-seven percent belong to an ethnic group other than Canada's White majority. It is one of the most diverse urban populations in the Western world, and it is home to one of North America's largest dialysis programs — annually serving over 6,000 patients with chronic kidney disease and more than 1,200 on dialysis.

Dr. Tabo Sikaneta is a nephrologist at the Scarborough Health Network. He has spent two decades treating kidney disease in this population, and in 2025 he published a study in *BMJ Open* that should change how every healthcare system thinks about the data it feeds its algorithms.

The headline finding was stark: within Canada's universal healthcare system, dialysis prevalence was 4.2 times higher among immigrants than non-immigrants. But Dr. Sikaneta did not stop at the headline. He disaggregated. He looked beneath the category of "immigrant" and found that the disparity was not uniform. It was concentrated in patients born in specific regions — the Caribbean, Southeast Asia, and South Asia. Kidney function declined fastest in patients from the Caribbean, South Asia, and East Asia. And the finding that forecloses the simplest explanations: *the year of immigration did not influence the outcomes.* A patient who arrived from the Caribbean thirty years ago showed the same disease trajectory as one who arrived five years ago.

This is not an acculturation problem. It is not a "new immigrant" problem. It is not, critically, a cost-access problem — because Canada's universal system removes the insurance barrier that distorts healthcare data in the United States. The disparity persists in a universal system, across decades of residency, concentrated in specific populations whose risk factors — genetic predisposition, epigenetic markers, environmental exposures in the country of origin, dietary and metabolic patterns that persist across generations — are not captured by any of the standard categories an algorithm would use.

An algorithm that sorted patients by "immigrant versus non-immigrant" would flag millions of healthy people as high risk while telling clinicians nothing about where to intervene. An algorithm that sorted by ethnicity would get closer but would still miss the birth-country specificity that Dr. Sikaneta's disaggregation revealed. Only a clinician with deep knowledge of the population — who knows which communities carry elevated risk, and that the risk persists regardless of how long the patient has lived in Canada — can look at a patient's chart and ask: is the category this system is using adequate to the complexity of this patient's risk?

That is the keeper function. And this chapter is about what happens when nobody performs it — when the algorithm is trusted, the categories are accepted, and the wrongness of the metric is invisible to everyone in the system.

The Wrong Metric

In the previous chapter, an algorithm encoded the biases of the criminal justice system and used them to predict who would commit a crime. The bias entered through the training data — arrest records, conviction rates, socioeconomic variables — that reflected a system of policing shaped by racial inequality. The output looked objective. The input was not.

This chapter tells a story that is, in some ways, more troubling. COMPAS was built by a private company selling a product to courtrooms. The algorithm at the center of this chapter was built by a health system trying to help its patients. The intentions were good. The designers were careful. The system was validated against its own metrics and performed well. And it was systematically denying care to Black patients who needed it, not because anyone instructed it to, but because the metric it was optimizing was the wrong metric — and the wrongness was invisible to everyone until a team of researchers decided to look.

The Algorithm

In 2019, Ziad Obermeyer, Brian Powers, Christine Vogeli, and Sendhil Mullainathan published a paper in *Science* that would reshape the conversation about algorithmic bias in healthcare. They examined a widely used commercial algorithm — deployed across hospitals and insurance systems to manage the care of approximately 200 million Americans — that was designed to identify patients who would benefit from enrollment in high-risk care management programs. These programs provide additional resources: dedicated nurse coordinators, more frequent check-ins, priority access to specialists. They are expensive and limited in capacity. The algorithm decided who got in.

The algorithm used a reasonable-sounding proxy for health need: healthcare costs. Patients who had spent more money on healthcare in the previous year were predicted to need more care in the following year. The logic was straightforward — if you spent a lot on healthcare last year, you are probably sick, and you will probably need a lot of healthcare next year. Cost as a proxy for need.

The algorithm worked. On its own terms, it performed well. It identified patients with high costs and predicted their future costs with reasonable accuracy. It was validated, deployed, and trusted by the institutions that used it.

There was one problem. The proxy was wrong.

The Gap

In the United States, healthcare spending is not determined solely by health need. It is determined by a complex interaction of insurance coverage, access to providers, geographic proximity to medical facilities, trust in the medical system, the ability to take time off work, transportation, language barriers, and — pervasively, measurably, and historically — race.

Black Americans, on average, spend less on healthcare than white Americans with comparable health conditions. This is not because Black Americans are healthier. It is because Black Americans face greater barriers to accessing care: lower rates of insurance coverage, fewer nearby providers, more experiences of discrimination in medical settings that discourage return visits, and the cumulative legacy of a medical system that, within living memory, experimented on Black patients without consent, denied Black patients treatment, and maintained segregated hospitals.

The reader who began this chapter in Scarborough will recognize the deeper problem. Dr. Sikaneta's research demonstrated that the disparity persists even when the cost barrier is removed — within universal healthcare, across decades of residency, concentrated in populations whose risk factors are not captured by the categories any algorithm would use. The Obermeyer algorithm's proxy was cost. But the Canadian evidence shows that the problem runs deeper than the proxy. The data itself carries the disparity. Fix the billing and you still have patients whose needs the standard categories understate.

The algorithm saw cost. The algorithm did not see the gap between cost and need. A Black patient who needed care but did not receive it — because they could not afford it, could not reach it, or did not trust the system enough to seek it — appeared in the data as a low-cost patient. Low cost, to the algorithm, meant low need. Low need meant the patient did not qualify for the care management program that might have addressed the very barriers preventing them from getting care in the first place.

The result, quantified by Obermeyer and his colleagues, was stark: at a given risk score, Black patients were considerably sicker than white patients with the same score. The researchers found that Black patients assigned the same level of predicted risk had 26% more chronic conditions than their white counterparts. The algorithm was not identifying the sickest patients. It was identifying the most expensive patients. And in a country where access to healthcare is unequal, the most expensive patients and the sickest patients are not the same people.

The researchers estimated that correcting the bias would increase the percentage of Black patients in the high-risk care management program from 17.7% to 46.5%. Nearly half of the Black patients who should have been receiving additional care were being denied it — not by a racist decision-maker, but by a metric that confused spending with suffering.

The Mirror

The COMPAS story and the healthcare story are structurally identical. In both cases, the algorithm used a proxy variable that correlated with the outcome of interest but also correlated with race, because the underlying system that produced the data was shaped by racial inequality. In COMPAS, the proxy was prior arrests (correlated with both reoffending and race through differential policing). In healthcare, the proxy was cost (correlated with both health need and race through differential access to care).

But the healthcare case is sharper in one specific respect. COMPAS was built by a private company selling a risk assessment tool to courtrooms. One can argue — and many have — that the commercial incentives were misaligned with justice. The healthcare algorithm was built by a health system trying to allocate scarce resources to the patients who needed them most. The intent was to help. The designers chose cost as a proxy because cost data is abundant, reliable, and easy to collect, while direct measures of health need — like the number and severity of chronic conditions — are harder to gather consistently at scale.

The choice was pragmatic and defensible. Cost data is clean. Health data is messy. The algorithm worked on the data it had. The problem was not carelessness or malice. The problem was that the data it had was a mirror of a society in which healthcare access is distributed unequally, and the mirror reflected the inequality as faithfully as it reflected everything else.

This is the characteristic that makes the Invisible Bias pattern so difficult to fight. The bias is not in the algorithm. The bias is in the world. The algorithm is a mirror. The mirror shows what is in front of it. If what is in front of it is a society where Black patients receive less care, the mirror will show Black patients needing less care. The mirror does not lie. But it does not tell the truth, either. It tells a version of the truth that is shaped by the angle at which it is held — and the angle, in this case, is determined by the metric the designers chose.

The Face

The same year the Obermeyer paper was published, a different study had already been making headlines for two years — one that demonstrated the same pattern in a domain where the stakes are different but the mechanism is identical.

In 2018, Joy Buolamwini, a researcher at the MIT Media Lab, and Timnit Gebru, then at Microsoft Research, published a paper called "Gender Shades." They evaluated three commercial facial recognition systems — from Microsoft, IBM, and Face++ — on their ability to correctly classify faces by gender. The results revealed a failure so systematic that it could not be attributed to random error.

For lighter-skinned males, the systems achieved error rates as low as 0.8%. For darker-skinned females, error rates reached 34.7%. The gap was not small. It was a factor of more than forty. The systems that technology companies were marketing as accurate and ready for deployment were, for a significant portion of the human population, barely functional.

The reason was the training data. Facial recognition systems learn to identify faces by being shown millions of photographs during training. The datasets that these systems were trained on were disproportionately composed of lighter-skinned faces, particularly lighter-skinned male faces. The systems became very good at recognizing the faces they saw most often and very bad at recognizing the faces they rarely saw. The bias was not in the algorithm's logic. It was in the algorithm's education — in the examples it was given to learn from.

The Deployment

A facial recognition system that fails 0.8% of the time for one group and 34.7% of the time for another is not merely inaccurate. When deployed in law enforcement — which it has been, across hundreds of police departments in the United States and governments around the world — it becomes a machine for generating false identifications that fall disproportionately on the people the system sees least clearly.

In 2020, Robert Williams, a Black man in Detroit, was arrested at his home in front of his wife and daughters for a theft he did not commit. The arrest was based on a facial recognition match that had misidentified him. Williams was held in jail for 30 hours before being released. The facial recognition system had matched his driver's license photo to surveillance footage of a different person. When the detective showed him the surveillance image, Williams held it next to his own face and said: "I hope you don't think all Black people look alike."

Williams was not the only case. In 2019, Nijeer Parks, a Black man in New Jersey, was arrested for shoplifting and attempting to hit a police officer with a car — at a location he had never been to, based entirely on a facial recognition match. He spent ten days in jail and nearly a year fighting the charges before they were dismissed.

In every documented case of a false facial recognition arrest in the United States, the person wrongly arrested has been Black. Every single one. This is not coincidence. It is the predictable consequence of deploying a system that is forty times more likely to misidentify dark-skinned faces in a context where misidentification leads to arrest, and arrest falls on the person the algorithm points to.

The Keeper in the Clinic

In the courtroom, the keeper is the judge. In the clinic, the keeper is the doctor.

A doctor who examines a patient, takes a history, runs tests, and makes a diagnosis is exercising judgment. The doctor can see things the algorithm cannot: the patient who minimizes their pain, the patient whose chart understates their condition because they have not been able to afford regular visits, the patient whose numbers look acceptable but whose face looks wrong. A doctor who knows their patient population — who understands the barriers to care that their patients face — can adjust. A doctor can say: the score says this patient is low risk, but I have seen enough patients from this community to know that low cost does not mean low need.

But the doctor, like the judge, faces the structural pressure of the number. The algorithm has been validated. It has been endorsed by the institution. It processes 200 million patients. The doctor who overrides the algorithm must justify the override. The doctor who follows the algorithm needs no justification. The path of least resistance is deference, and deference to a biased output reproduces the bias — at scale, automatically, invisibly.

This is the pattern from Chapter 4, translated again. The autopilot degrades the pilot's skill. COMPAS degrades the judge's authority. The healthcare algorithm degrades the doctor's clinical judgment. In each case, the tool that was designed to assist the keeper becomes the mechanism by which the keeper's judgment is displaced. The keeper remains in the room. The keeper's role shrinks.

But the keeper who opened this chapter did not shrink. Dr. Sikaneta looked at the same data the algorithm would have used and asked whether the categories were adequate. He disaggregated when the system would have aggregated. He found the birth-country specificity that no standard classification captures. He demonstrated that the risk persists across decades in a universal system — a finding that changes clinical practice for the communities he serves. That is what it looks like when the keeper's role does not shrink. It is also, in the current trajectory of healthcare AI, the exception rather than the rule.

The Compounding

There is a feature of algorithmic bias in healthcare that does not have an equivalent in the courtroom, and it is worth naming because it is the mechanism by which the harm deepens over time.

When the healthcare algorithm classifies a Black patient as low risk and denies them entry to the care management program, the patient does not receive the additional care. Because the patient does not receive the additional care, the patient's health may deteriorate. But the deterioration does not necessarily increase the patient's healthcare spending — the patient may still face the same barriers to care that kept their spending low in the first place. The patient gets sicker. The spending stays low. The algorithm, in the next cycle, still classifies the patient as low risk.

The bias compounds. The algorithm's output — which denies care — produces the data — low spending — that the algorithm uses to justify denying care in the next cycle. The system generates the evidence for its own correctness. A patient who is denied care looks, in the data, like a patient who does not need care. The denial becomes invisible. It looks like health.

This is a feedback loop, and it is one of the most dangerous features of algorithmic systems deployed in consequential domains. The algorithm does not merely reflect the past. It shapes the future in ways that confirm its own assumptions. The criminal justice version of this loop is well documented: predictive policing algorithms direct more officers to neighborhoods flagged as high crime, which produces more arrests in those neighborhoods, which produces more data showing those neighborhoods are high crime, which justifies directing more officers there. The healthcare version is quieter but equally corrosive: the algorithm that denies care to the undertreated produces data showing the undertreated are low need, which justifies continuing to deny them care.

The Response That Worked

The Obermeyer paper is unusual in the landscape of algorithmic bias research because it has a relatively hopeful coda. The researchers worked with the algorithm's developer, and the company agreed to change the metric. Instead of predicting cost, the revised algorithm predicts a combination of health indicators that more directly measure need. The bias was reduced by approximately 84%.

This matters because it demonstrates that the problem is not insoluble. The algorithm was not inherently racist. It was optimizing the wrong thing. Change the metric and the bias decreases. This is evidence that careful, critical examination of algorithmic systems can identify and correct harms.

But the correction happened only because a team of researchers chose to investigate. For the years before the paper was published, the algorithm was operating as designed, endorsed by the institutions that used it, processing 200 million patients, and systematically underserving Black patients — and nobody inside the system detected the problem. The keeper — in this case, the researchers who performed the audit — was not built into the system. The keeper arrived from outside, years later, and only because academic researchers had the motivation, the access, and the statistical tools to look.

There is a tension here that this chapter must name. The algorithm that encoded the bias was also the instrument that made the bias visible, measurable, and correctable — a function the human system had failed to perform for decades. Before the algorithm, human gatekeepers in the healthcare system were also failing Black patients — that is why spending was lower in the first place. The disparity was real and the humans in the system did not detect it. The algorithm did not create the injustice. It inherited the injustice and, by encoding it consistently, made it examinable for the first time. The lesson is not that algorithms are biased. The lesson is that algorithms inherit the biases of the systems they learn from — and the correction requires a keeper who can recognize the bias once it is exposed.

How many algorithms are running today, at similar scale, with similar biases, that have not yet been audited? The answer is not known. The question is not rhetorical. It is the central policy question of algorithmic governance, and the answer is almost certainly: many.

The Architecture of Invisible Harm

The Therac-25 killed three patients. The families knew. The hospitals knew. The deaths were investigated, the machine was recalled, and the flaw was identified. The harm was visible.

The healthcare algorithm denied additional care to tens of thousands of Black patients over multiple years. Nobody knew. No alarm sounded. No patient was told: you were denied this program because the algorithm scored you as low risk, and the algorithm scored you as low risk because you are Black and Black patients spend less on healthcare and the algorithm confused cost with need. The harm was invisible — not because it was small, but because it was distributed, statistical, and indistinguishable, at the individual level, from the ordinary workings of a healthcare system that already underserves Black patients.

This is the architecture of invisible harm, and it is the architecture of the remaining chapters. The sentencing machine produces bias that is invisible because it is dissolved into a number. The diagnostic produces bias that is invisible because it looks like health. The feed, in the next chapter, produces harm that is invisible because it looks like engagement — people clicking, sharing, commenting — while the content they are engaging with is driving them toward hatred and violence.

The common thread: in each case, the system's own metrics say it is working. The risk score predicts cost accurately. The facial recognition system identifies most faces correctly. The engagement metrics are up. The system is succeeding by its own definition of success. The harm is in the gap between the system's definition of success and the human definition of success — and that gap is invisible to the system, invisible to its operators, and visible only to the people who fall into it.

> *"The system is succeeding by its own definition of success. The harm is in the gap between the system's definition of success and the human definition. That gap is visible only to the people who fall into it."*

The sentencing machine misread futures. The diagnostic misread the present. The next chapter tells the story of a system that did not misread anything at all. It read the data perfectly, optimized its objective function precisely, and achieved its goal exactly — and the goal, measured in clicks and shares and time on screen, was perfectly aligned with the amplification of content that drove a population toward genocide.

The COMPAS algorithm was biased by accident. The healthcare algorithm was biased by proxy. The recommendation engine at the center of the next chapter was not biased at all. It was optimized. And the optimization produced mass atrocity.

The Last Keeper

The Feed

On the Facebook recommendation engine, the UN finding that a
social media platform played a "determining role" in inciting
genocide, and the discovery that a system optimized for engagement
will always amplify the content that provokes the strongest reaction

"The algorithm does not know what hate is.
It knows what engagement is.
In practice, they are the same signal."

The Feed

On the Facebook recommendation engine, genocide, and the discovery that engagement optimization is indistinguishable from radicalization

The previous two chapters told stories about algorithms that encoded bias by accident. COMPAS absorbed the racial disparities in the criminal justice system through its training data. The healthcare algorithm absorbed the racial disparities in healthcare access through its choice of metric. In both cases, the designers did not intend the harm. The harm was a side effect of a reasonable design choice interacting with an unequal world.

This chapter tells a different story. The algorithm at the center of this chapter was not biased. It was not making an error. It was not confused about its metric. It was optimizing, with extraordinary effectiveness, for the objective it was given: engagement. The problem is that engagement, measured at the scale of billions of interactions per day, is a metric that does not distinguish between a user clicking on a friend's birthday photo and a user clicking on content that tells them their neighbors are subhuman and should be exterminated. Both are engagement. Both generate the signal the algorithm is trained to maximize. And the content that provokes the most intense emotional reaction — which is, reliably, content that divides, enrages, and dehumanizes — generates more engagement than the content that informs, connects, or entertains.

COMPAS was biased by accident. The healthcare algorithm was biased by proxy. The recommendation engine was not biased at all. It was optimized. And the optimization produced mass atrocity.

The Country

Myanmar is a country of approximately 54 million people in Southeast Asia, bordered by India, China, Bangladesh, Laos, and Thailand. For decades, it was governed by a military junta that restricted press freedom, political opposition, and access to information technology. When the country began opening politically and economically in the early 2010s, internet access expanded rapidly — from roughly 1% of the population in 2010 to over 30% by 2016, driven almost entirely by inexpensive smartphones and mobile data plans.

For most people in Myanmar, Facebook was the internet. This is not a metaphor. Mobile carriers sold data plans that included free or discounted access to Facebook. Many users did not distinguish between Facebook and the internet itself. The platform was where people read news, communicated with family, conducted business, and encountered information about the world beyond their immediate community. Facebook had approximately 20 million users in Myanmar by 2018 — roughly one-third of the country's population.

The Rohingya are a Muslim ethnic minority living predominantly in Rakhine State, in Myanmar's west. They have been subject to systematic discrimination by the Buddhist-majority state for decades — denied citizenship under a 1982 law, restricted in movement, education, and marriage, and subjected to periodic waves of communal violence. The tensions between the Rohingya and the ethnic Rakhine Buddhist population are deep, historical, and complex.

What happened on Facebook was not complex. It was an amplification machine pointed at a population primed for violence, with no one watching the output.

The Content

Beginning in the mid-2010s, anti-Rohingya content proliferated on Facebook in Myanmar. Nationalist Buddhist monks, military-linked accounts, and coordinated networks of fake profiles posted and shared content that dehumanized the Rohingya: fabricated stories of Rohingya violence against Buddhists, manipulated images purporting to show Muslim atrocities, calls for the expulsion or elimination of the Rohingya population, and inflammatory rhetoric that described the Rohingya as animals, invaders, and threats to Buddhist identity.

The content was in Burmese. Facebook, at the time, had approximately two dozen content moderators who spoke Burmese — for a platform serving 20 million users. The moderation was not merely inadequate. It was, at the scale required, impossible. A content moderation team of two dozen people cannot review the output of 20 million users producing content in a language the company's automated systems did not reliably understand. The content was, in effect, unmoderated.

But it was not unamplified.

Facebook's recommendation algorithm did not create the anti-Rohingya content. But it distributed it. The algorithm observed that users who saw inflammatory anti-Rohingya posts engaged with them at high rates — clicking, commenting, sharing, reacting. The algorithm interpreted this engagement as a signal of user interest and showed more similar content to those users and to users with similar profiles. The content that generated the most engagement was, predictably, the content with the strongest emotional charge: the most extreme, the most frightening, the most dehumanizing.

The feedback loop from the previous chapter operates here at planetary scale. The algorithm shows inflammatory content. Users engage with inflammatory content. The algorithm sees the engagement and shows more inflammatory content. Users engage more. The cycle accelerates. The feed becomes progressively more extreme — not because anyone decided to radicalize the users, but because radicalization is what engagement optimization looks like when the content involves an ethnic minority that a significant portion of the population already distrusts.

The Violence

In August 2017, the Myanmar military launched what it called "clearance operations" in Rakhine State. Over the following months, security forces and allied militias burned Rohingya villages, committed mass killings, and carried out systematic sexual violence. Approximately 700,000 Rohingya fled across the border to Bangladesh, creating one of the largest refugee crises in modern history. The UN High Commissioner for Human Rights described the operations as "a textbook example of ethnic cleansing." Multiple international bodies, including the International Court of Justice, have since characterized the violence as genocide.

In 2018, the UN Human Rights Council's Independent International Fact-Finding Mission on Myanmar published its report. The mission had been established to investigate the human rights situation in Myanmar, with particular focus on the Rohingya crisis. The report's findings regarding Facebook were unambiguous.

The mission found that Facebook had been used to "spread hate, incite violence, and coordinate attacks" against the Rohingya. The chairman of the fact-finding mission, Marzuki Darusman, stated that Facebook had played a "determining role" in the crisis. The report documented specific instances in which content posted on Facebook — fabricated stories of Rohingya attacks, calls to violence by military-linked accounts, dehumanizing rhetoric from nationalist organizations — had preceded and accompanied specific acts of physical violence.

The Optimization

It would be comforting to describe Facebook's role in the Myanmar genocide as a failure. A bug. A system that malfunctioned. But that description is not accurate.

The recommendation algorithm worked. It identified content that users engaged with. It showed users more content like it. It increased time on platform, increased interactions, increased the metrics that Facebook's business model depends on. By every measure internal to the system, the algorithm was succeeding.

The problem was not that the system failed to optimize. The problem was what optimization means when the objective is engagement and the context is a society on the verge of mass violence. Engagement optimization does not distinguish between engagement driven by curiosity and engagement driven by fear. It does not distinguish between a user who clicks because they want to learn and a user who clicks because they are being radicalized. It does not distinguish between content that builds community and content that incites genocide. All of these produce clicks. All of these produce shares. All of these produce comments. All of these are, to the algorithm, the same signal.

This is the chapter's central insight, and it connects to every chapter that precedes it: **the danger of an automated system is not always that it fails. Sometimes the danger is that it succeeds at the wrong objective.** The Therac-25 failed to check the beam configuration. The autopilot failed and handed off to deskilled pilots. MCAS succeeded at pushing the nose down but the sensor was wrong. COMPAS succeeded at predicting cost but cost was a proxy for race. The healthcare algorithm succeeded at identifying high-spending patients but spending was a proxy for access. And the recommendation engine succeeded at maximizing engagement but engagement was a proxy for outrage.

The Speed

The Myanmar case also demonstrates Pattern 4 — the Speed-Judgment Tradeoff — at a scale that dwarfs any previous chapter.

Facebook processes billions of content interactions per day across its platforms. The recommendation algorithm operates continuously, in real time, evaluating content, predicting engagement, and distributing posts to feeds at a rate that no human moderation system can match. The two dozen Burmese-speaking moderators were not inadequate because they were insufficient in number — though they were. They were inadequate because the task itself is impossible at the speed the algorithm operates. By the time a human moderator can evaluate a piece of content, the algorithm has already shown it to thousands or millions of users. The content has already generated engagement. It has already been shared. It has already been absorbed.

In the Therac-25, the operator had seconds. In Air France 447, the pilots had three minutes. In the 737 MAX, the pilots had eleven minutes. On Facebook, the window for human intervention is measured in milliseconds — the time between the algorithm's ranking decision and the content's appearance on a user's screen. The keeper cannot operate at this speed. No human can. The system has accelerated past the point where human judgment can intervene before the consequence.

This is the Speed-Judgment Tradeoff at its endpoint: a system so fast that inserting a human into the loop would require slowing the system to a pace incompatible with its function. Facebook's recommendation algorithm works because it is instant. Making it slow enough for human review would make it something other than what it is. The speed is not a feature of the system. The speed is the system.

The Warnings That Were Ignored

Facebook was warned. This is documented.

As early as 2013, civil society organizations in Myanmar raised concerns about the spread of anti-Muslim hate speech on the platform. Researchers, journalists, and human rights groups contacted Facebook repeatedly, identified specific accounts and networks spreading inflammatory content, and requested action. In 2014, a group of technology researchers published a detailed report on the role of Facebook in amplifying communal tensions in Myanmar.

Facebook's response was slow, inadequate, and, for the most consequential period, effectively absent. The company did not invest in Burmese-language content moderation at a scale remotely proportional to the platform's influence in the country. It did not prioritize the development of automated tools capable of detecting hate speech in Burmese. It did not establish local partnerships sufficient to monitor the platform's effects on the ground. The warnings were received. The warnings were not acted upon with the urgency the situation required.

In 2018, after the genocide was already underway, Facebook's own internal assessment acknowledged that the company had not done enough. Mark Zuckerberg told Congress that the situation in Myanmar was "a terrible tragedy" and that the company was "too slow to act." In a subsequent interview, he acknowledged that Facebook needed to "do more" but resisted the characterization that the platform bore significant responsibility for the violence.

The pattern is, by now, familiar to the reader. AECL was told the Therac-25 was harming patients and could not reproduce the error. Boeing was told by its own test pilots that MCAS was problematic and the schedule held. Facebook was told that its platform was being used to incite violence against an ethnic minority and did not invest in the moderation required to address it. In every case, the institution received the information and the institution did not act with the urgency the information demanded. The institution's confidence in its system — its engineering, its product, its platform — exceeded its willingness to believe that the system could produce catastrophic harm.

The Question of the Keeper

Where is the keeper in a recommendation engine?

In the Therac-25, the keeper was a hardware interlock. In the cockpit, the keeper was the pilot. In the courtroom, the keeper was the judge. In the clinic, the keeper was the doctor. In each case, the keeper was identifiable — a specific mechanism or a specific person positioned at the point where the system's output meets the human who is affected by it.

A recommendation engine has no such point. There is no single moment where a human intervenes between the algorithm's decision and the user's exposure to the content. There are billions of such moments, every day, and each one is automated. The keeper would need to be present at every feed refresh, every recommendation, every ranking decision — and would need to evaluate each one for its potential contribution to radicalization, communal violence, or genocide in a specific cultural context. This is not a task a human can perform. It is not a task a thousand humans can perform. The system has scaled past the point where the keeper model, as defined in this book's opening chapter, can function.

What happened instead is worse than the keeper's absence. The burden was silently transferred. Because the platform outpaced its human moderators — fewer than two dozen Burmese-speaking reviewers for twenty million users — the responsibility for distinguishing truth from incitement was offloaded onto the end user. Farmers in rural Myanmar, many of whom had only recently acquired smartphones and for whom Facebook *was* the internet, were implicitly asked to be the keepers of their own reality: to evaluate algorithmic recommendations for genocidal content in real time, without training, without context, and without any indication that what they were seeing had been selected by a machine optimising for engagement rather than accuracy. Human psychology is not equipped for this task. We are pattern-matching animals wired to trust the information our environment presents to us. The algorithm exploited exactly that wiring. The keeper was not removed. The keeper was made into the victim — given responsibility without capacity, accountability without tools, and a feed that looked like the world but was, in fact, a machine's guess at what would keep them scrolling.

And that is precisely the point. The book began with a lighthouse keeper — one person, one light, one stretch of coast. Each subsequent chapter has expanded the scope: one treatment room, one cockpit, one courtroom, one clinic. The recommendation engine serves two billion users. The question is no longer "where is the keeper?" The question is whether the concept of a keeper has meaning at a scale where no human judgment can touch the system's output before it reaches the people it affects.

The Diffusion of Accountability

Pattern 5 — the Accountability Gap — reaches its most extreme expression in this chapter.

When the Therac-25 killed patients, AECL was identifiable as the responsible party. When the 737 MAX crashed, Boeing was indicted. Responsibility, however imperfectly assigned, could be traced.

Who is responsible for the Myanmar genocide? The military commanders who ordered the operations bear primary responsibility. The soldiers who carried out the killings bear responsibility. The political leaders who enabled the military bear responsibility. But what of the platform that amplified the content that dehumanized the victims? What of the algorithm that ranked that content for maximum visibility? What of the company that designed the algorithm, deployed it in a country it did not understand, ignored warnings that the platform was being used as a weapon, and did not invest in the moderation that might have blunted the amplification?

Facebook did not order the genocide. Facebook did not intend the genocide. Facebook's algorithm did not know what the Rohingya were, what genocide is, or what the content it was amplifying would lead to. The algorithm knew that certain content generated more engagement. It showed more of that content. The content happened to be calls for ethnic cleansing.

The accountability gap here is not a matter of blame being diffused across a complex system, as in the 737 MAX. It is a matter of a new kind of causation — algorithmic amplification — for which existing legal and moral frameworks have no adequate vocabulary. Facebook is not an author of the content. It is not a publisher in the traditional sense. It is not a broadcaster. It is a distribution system that amplifies content based on predicted engagement, without evaluating the content's truth, its social impact, or its potential to contribute to violence. There is no precedent, in law or in ethics, for this kind of entity — and the absence of precedent is itself a form of impunity.

The Algorithmic Judge: A Reckoning

The last three chapters have traced a progression.

In Chapter 7, the algorithm encoded bias by accident — absorbing the racial disparities of the criminal justice system through its training data and producing risk scores that were used to keep human beings in cages. The harm was in the bias. The keeper — the judge — was present but deauthorized.

In Chapter 8, the algorithm encoded bias by proxy — using cost as a stand-in for need in a country where cost and need diverge along racial lines. The harm was in the metric. The keeper — the doctor — was present but displaced.

In this chapter, the algorithm was not biased at all. It optimized its objective function with precision. The harm was in the objective itself — engagement, measured without reference to truth, safety, or human consequence. The keeper — the content moderator — was absent at scale.

Accident. Proxy. Optimization. Each step is harder to detect and harder to fight than the last. Accidental bias can be identified through audit and corrected through redesign. Proxy bias can be identified through careful analysis of the metric and corrected by choosing a better one. But optimization bias — a system that harms people precisely *because* it is achieving its goal — cannot be corrected by making the system work better. It can only be corrected by changing the goal. And changing the goal of a system that generates hundreds of billions of dollars in revenue requires a force external to the system: regulation, public pressure, or the kind of institutional conscience that the next chapters will test.

> *"Accidental bias can be audited and fixed. Proxy bias can be identified and the metric changed. But optimization bias — a system that harms because it succeeds — can only be corrected by changing the goal. And the goal generates hundreds of billions in revenue."*

From the Feed to the Battlefield

Part III asked what happens when algorithms judge. Part IV asks what happens when algorithms act — in the most consequential domains remaining: the selection of military targets, the surveillance of entire populations, and the engineering systems whose complexity exceeds human capacity to verify.

The recommendation engine amplified content that contributed to genocide. The next chapter describes systems designed explicitly for the purpose that the recommendation engine served incidentally: identifying human targets. The AI targeting systems described in the next chapter do not amplify calls for violence. They are the violence — automated, accelerated, and operating at a speed that reduces the human in the loop to a formality.

The Last Keeper

The Target

On AI-assisted military targeting, the machines that generate
kill lists faster than humans can review them, and the moment
when the keeper in the loop becomes a signature on a form
the machine has already decided

*"A human in the loop who has twenty seconds to approve a target
generated by a machine that has spent hours analyzing the data
is not exercising judgment. They are providing legitimacy."*

The Target

On July 3, 1988, the USS Vincennes, an American guided-missile cruiser operating in the Strait of Hormuz during the Iran–Iraq War, fired two surface-to-air missiles at an aircraft climbing away from Bandar Abbas International Airport. The aircraft was Iran Air Flight 655, a civilian Airbus A300 on a scheduled commercial route to Dubai. All 290 people aboard — including 66 children — were killed.

The Vincennes was equipped with the Aegis Combat System, the most advanced naval warfare computer of its era. The Aegis system tracked the approaching aircraft, classified it, and presented data to the ship's combat information center. The crew, operating under the stress of an ongoing surface engagement with Iranian gunboats, interpreted the data as indicating a descending Iranian F-14 fighter on an attack profile. The aircraft was in fact ascending, on a standard commercial departure trajectory, broadcasting a civilian transponder signal.

The subsequent investigation found that the Aegis system had correctly identified the aircraft's altitude, speed, and transponder signal. The data was on the screens. The crew misread the data, or selectively attended to the data that confirmed their expectation of an attack, or were overwhelmed by the volume of information the system presented during a high-stress engagement. The system provided the information. The humans made the wrong decision.

The Vincennes incident is the historical anchor for this chapter because it demonstrates, in 1988, the central tension that AI-assisted targeting amplifies thirty-five years later: **a human presented with machine-generated data, under time pressure, in conditions of stress, will tend to defer to the interpretation that the system's presentation makes most salient — even when the data, correctly read, contradicts that interpretation**. The keeper was in the loop. The keeper had the information. The keeper killed 290 civilians.

> *"The data was on the screens. The aircraft was ascending, broadcasting a civilian signal. The crew fired anyway. The keeper was in the loop. The keeper had the information. The keeper killed 290 civilians."*

The Acceleration

The Vincennes crew had approximately three minutes between initial detection and the decision to fire. Three minutes to identify the aircraft, assess the threat, request authorization, and launch. In that window, the crew processed data from the Aegis system, communicated with higher command, and reached a conclusion that was wrong. Three minutes was not enough time for the human judgment required.

Modern AI-assisted targeting systems operate at a fundamentally different tempo. The systems reported on by investigative journalists and described by military sources do not present data to a human and wait for a decision. They process vast quantities of surveillance data — signals intelligence, communications intercepts, movement patterns, social network analysis, geolocation data — and produce targeting recommendations. Not raw data for a human to interpret. Recommendations. Ranked lists. Predicted identities. Confidence scores.

The shift from data presentation to recommendation is the shift that transforms the human's role. When the Aegis system showed the Vincennes crew a radar track, the crew had to interpret it: is this track a threat? The interpretation was wrong, but it was a human act of judgment. When an AI targeting system generates a list of recommended targets with confidence scores, the human's role is not interpretation. It is approval. The machine has already judged. The human is asked to ratify the machine's judgment.

This is the COMPAS pattern translated to the battlefield. The judge sees a risk score and is structurally pressured to defer. The military operator sees a targeting recommendation and is structurally pressured to approve. The path of least resistance is deference — because overriding the machine requires justification, and the machine has already done the analysis that the human would need to perform independently to reach a contrary conclusion. The human does not have the time, the data access, or the analytical capacity to replicate the machine's work. The human can approve or reject. The human cannot meaningfully evaluate.

The Reports

In 2024, the Israeli-Palestinian investigative outlet +972 Magazine, in collaboration with the Hebrew-language publication Local Call, published a series of investigative reports based on interviews with Israeli intelligence officers describing AI-assisted targeting systems used during the military operations in Gaza. The reports described two systems in particular.

The first, referred to as "Gospel," was described as a system that generates recommendations for military targets — buildings, infrastructure, and sites assessed to be linked to militant activity. Sources described Gospel as producing targeting recommendations at a rate that dramatically exceeded the capacity of previous intelligence processes. Where earlier methods might produce a limited number of vetted targets over weeks, Gospel was described as generating hundreds of targets in compressed timeframes.

The second, referred to as "Lavender," was described as a system that identifies individuals assessed to be associated with militant organizations. Sources described Lavender as processing surveillance data to generate lists of individuals ranked by their predicted association with militant groups, with each individual assigned a numerical score. Sources stated that the system's output was treated, in practice, as equivalent to a targeting decision — that the human review of individual targets generated by Lavender was minimal, sometimes lasting as little as seconds per target.

Twenty seconds. To review a machine's recommendation to kill a human being.

The reporting described a process in which the AI system performed the substantive analytical work — correlating data sources, assessing patterns of life, estimating association with militant networks — and the human performed a nominal review that one source compared to a formality. The system generated the list. The human approved the list. The strike was executed. Often, according to the reporting, strikes were carried out on targets while they were in their homes, where they were surrounded by family members, because targeting them in their homes was operationally simpler than targeting them in other locations.

The Rubber Stamp

The concept of a "human in the loop" is central to the international legal and ethical framework governing the use of lethal force. International humanitarian law requires that attacks distinguish between combatants and civilians, that the anticipated military advantage be proportional to the expected civilian harm, and that precautions be taken to minimize civilian casualties. These obligations presuppose a decision-maker — a human who evaluates the target, assesses proportionality, considers the civilian environment, and makes a judgment about whether the strike is lawful.

A human who has twenty seconds to review a machine-generated target is not performing this function. Twenty seconds is not enough time to independently assess the intelligence that produced the targeting recommendation. It is not enough time to evaluate the civilian environment around the target. It is not enough time to consider proportionality in any meaningful sense. Twenty seconds is enough time to look at a name, see a confidence score, and approve.

This is the automation paradox's most lethal expression. The system generates targets so efficiently that the human review process cannot keep pace without being compressed to the point of meaninglessness. The more targets the system produces, the less time the human has per target. The less time the human has, the less meaningful the review. The less meaningful the review, the more the system's output becomes, in practice, the final decision.

The human is in the loop. The human has not been removed. But the human's role has been compressed to the point where "in the loop" and "out of the loop" are functionally indistinguishable. The keeper is present. The keeper is not keeping.

The Collateral Arithmetic

The +972 reporting described another dimension of the AI targeting process: the management of expected civilian casualties. Sources described a system in which each targeting recommendation was accompanied by an estimated number of civilians likely to be killed in the strike. This estimate was generated by the system based on data about the target's known location, the building's occupancy patterns, and the weapon selected for the strike.

Sources described a threshold: a pre-authorized number of civilian casualties deemed acceptable for targets of different ranks or assessed importance. For lower-ranking targets, the acceptable number was described as relatively low. For higher-ranking targets, the acceptable number was described as significantly higher. The system generated the target. The system estimated the civilian cost. The human compared the estimate to the threshold. If the number was below the threshold, the strike was approved.

The arithmetic matters. The system is not making a binary decision — strike or don't strike. It is making a calculation: is the expected number of dead civilians acceptable given the assessed importance of the target? And the target's importance was assessed by the same AI system that generated the recommendation. The machine determines who the target is. The machine determines how important the target is. The machine estimates how many civilians will die. The human compares two numbers and authorizes.

Every step in this chain that requires judgment — is this person a combatant, how important are they, how many civilian deaths are acceptable — has been delegated to the machine or reduced to a numerical threshold. The human's judgment has been replaced by the human's arithmetic. And the arithmetic, as we saw in Chapter 6, is easier to perform than the judgment — which is precisely why the system is designed this way.

The Legal Void

International efforts to regulate autonomous weapons have been underway for over a decade. The Convention on Certain Conventional Weapons (CCW), a UN framework, has hosted discussions on lethal autonomous weapons systems since 2014. Multiple countries and civil society organizations have called for a binding international treaty. As of this writing, no such treaty exists.

The core difficulty is definitional. What counts as an autonomous weapon? A fully autonomous system that selects and engages targets without any human involvement is one end of a spectrum. An AI system that generates targeting recommendations reviewed by a human for twenty seconds is somewhere in the middle. A precision-guided munition that tracks a designated target after launch is at the other end. The line between "autonomous" and "assisted" is not bright. It is a gradient, and the gradient is defined by the quality of human involvement — a quality that, as this chapter has argued, can be degraded to the point of meaninglessness while remaining technically present.

This is the accountability gap at its widest. If a soldier fires a rifle and kills a civilian, the chain of responsibility is clear. If an AI system generates a target, a human approves it in twenty seconds, a drone strikes a building, and civilians die, who is responsible? The system designer who built the algorithm? The intelligence analyst who trained it? The operator who approved the target? The commander who set the acceptable casualty threshold? The political leadership that authorized the use of the system?

The answer, in practice, has been: all of them, and therefore none of them. The responsibility diffuses. The system absorbs it. And the civilians who die in the strike are recorded as collateral damage within acceptable parameters — parameters set by the same system that generated the target.

The Red Lines

In February 2026, an event occurred that connected this chapter directly to the thesis of this book.

Anthropic, the company that builds the Claude AI system, was offered a contract with the US Department of Defense valued at approximately $200 million. The company's leadership evaluated the contract and identified two applications that it determined were incompatible with its values: the use of its AI for autonomous weapons targeting and for mass surveillance of civilian populations. Anthropic declined those specific applications while remaining open to other defense work.

The response was immediate and severe. Elements within the defense establishment pushed to designate Anthropic as a supply-chain risk — a classification that would effectively blacklist the company from government contracts. The argument was that a company unwilling to provide unrestricted AI capabilities to the military could not be relied upon as a strategic partner.

The public responded differently. Within weeks of the confrontation becoming public, Claude became the number one application in the App Store. The company reported over one million daily sign-ups. The growth was driven predominantly by users who had never previously used the product — people who chose Anthropic specifically because the company had drawn a line.

This is the keeper's last stand in the AI era, and it is worth naming precisely. Anthropic did not argue that AI should not be used by the military. It did not refuse all defense work. It drew two specific lines: no autonomous weapons, no mass surveillance. These are the two applications where the keeper is most obviously removed from the loop — where the machine selects targets without meaningful human judgment, and where the machine monitors populations without individualized human review. The company that builds the AI said: these two uses remove the human from loops where the human must remain.

Whether this line holds is an open question. The economic and political pressures to cross it are enormous. The precedent, as this entire book has documented, is that lines drawn against the removal of the keeper are eventually crossed — because the machine is faster, because the machine is cheaper, because the machine is more consistent, and because the people who benefit from the machine's efficiency do not bear the cost of the machine's errors.

The Pattern at Its Endpoint

This chapter is the convergence point of the book's five patterns.

The **Reliability Trap**: the more effective AI targeting becomes, the more dependent military operations become on it, the less capable human analysts become of performing the targeting function independently, and the worse the consequences when the system is wrong.

The **Single Point of Trust**: the system's targeting recommendation is treated as equivalent to a verified intelligence assessment. One system. One score. One decision.

The **Invisible Bias**: the system is trained on data from previous conflicts, previous intelligence assessments, previous targeting decisions — all of which contain the biases and errors of those prior operations. The patterns the system learns include the patterns of who was targeted before, which may not correspond to who should be targeted now.

The **Speed-Judgment Tradeoff**: the system generates targets faster than humans can review them. The human review is compressed to the point of formality. The speed of the system is the mechanism that removes the human from the loop while maintaining the appearance of human control.

The **Accountability Gap**: when a strike kills civilians, responsibility diffuses across the system — the algorithm, the operator, the commander, the threshold, the policy. Nobody is responsible because everyone is responsible. The system absorbs the accountability that should attach to a human decision to take a human life.

The Doctrine

As this book was being completed, the targeting systems described in this chapter stopped being experimental.

On February 28, 2026, the United States and Israel launched Operation Epic Fury — a massive joint military campaign against Iran that struck over 1,000 targets within the first 24 hours. Artificial intelligence tools were deployed in active targeting and intelligence support roles. The kinetic strikes were accompanied by AI-enabled cyber warfare, including the hacking of a widely used religious calendar app to deliver targeted psychological messaging to over five million users. What was experimental in Gaza — Lavender, The Gospel, the AI targeting systems that generated lists of human beings for lethal action — had become standard operating doctrine for a major superpower conflict in less than two years.

I want to say something about this that the operational language of military planning is designed to obscure.

When a targeting system calculates that a strike on a building will kill twelve civilians and three combatants, and the operator approves the strike based on that calculation, the twelve civilian deaths are not incidental. They are not the fog of war. They are not imperfect intelligence. They were computed in advance. The system modeled their deaths before they happened, assigned their lives a numerical weight, measured that weight against a military objective, found the weight acceptable, and proceeded. Those twelve people were dead in the model before the missile launched.

That is not collateral damage. That is planned civilian killing with a mathematical justification. International humanitarian law has always acknowledged that civilians die in war. The legal and ethical framework handles this through the doctrine of proportionality — the *unintended* civilian harm must not be excessive relative to the military advantage. The key word is unintended. But when a system computes the civilian deaths in advance, presents them to an operator as an acceptable ratio, and the operator approves — those deaths are not unintended. They are planned. They were modeled, calculated, weighed, and accepted before the weapon was fired.

The mathematical justification is what makes it feel clean. A soldier who shoots a child experiences moral injury. A system that outputs "12 estimated civilian casualties, acceptable under current engagement parameters" experiences nothing. The moral weight evaporates into the calculation. The number creates distance between the decision and its consequence. And the person who approves the number — in twenty seconds, as the reporting on Lavender documented — is not making a judgment. They are ratifying a computation. The human in the loop is not exercising discernment. They are providing a signature.

> *"When the system models civilian deaths in advance, assigns their lives a numerical weight, and proceeds — those deaths are not collateral damage. They are planned. The number creates distance between the decision and its consequence, and the moral weight evaporates into the calculation."*

The speed of the transition — from experimental deployment to standard doctrine in less than two years — is itself the warning. The governance frameworks that might have constrained these systems were not built in time. The window between "experimental" and "doctrine" closed before anyone outside the military could meaningfully intervene. That is the pattern this book has traced from its first page: the machine moves faster than the institutions charged with governing it.

From the Target to the Watcher

AI targeting selects individuals for lethal action. The next chapter describes the infrastructure that makes targeting possible — the surveillance systems that collect the data the targeting algorithms consume. Mass surveillance is not a separate issue from autonomous weapons. It is the upstream infrastructure: the sensor network that feeds the targeting system, the monitoring apparatus that identifies the individuals the algorithm will assess, and the mechanism by which entire populations are reduced to data points in a system optimized for a purpose those populations did not consent to and cannot escape.

The Last Keeper

The Watcher

On mass surveillance infrastructure, the systems that monitor
entire populations, and the discovery that the watcher is where
all five of the book's patterns converge simultaneously

*"The question is not whether you are being watched.
The question is whether anyone is watching the watcher."*

The Watcher

On mass surveillance infrastructure, the systems that monitor entire populations, and the keeper who watches everyone except itself

In 2011, Malte Spitz, a German politician and privacy advocate, sued his mobile phone carrier, Deutsche Telekom, to obtain the records the company had collected about him under Germany's mandatory data retention law. After a legal battle, Deutsche Telekom released the records. Spitz had them visualized by the German newspaper *Die Zeit*.

The dataset covered six months. It contained 35,831 data points. Each point recorded Spitz's location, the time, and whether he was making a call, sending a text, or using mobile data. Plotted on a map and animated over time, the data produced a detailed reconstruction of Spitz's life: where he slept, where he worked, when he traveled, which routes he took, whom he visited, how long he stayed. The visualization showed not just where Spitz was at any given moment but the pattern of his existence — the rhythms, the habits, the relationships made visible through proximity and frequency.

That was 2011. One person. One carrier. Six months. 35,831 data points collected as a byproduct of normal phone use.

The surveillance infrastructure described in this chapter does not collect data as a byproduct. It is purpose-built. It operates at the scale of entire populations. And it is augmented by artificial intelligence that does not merely record data but interprets it — identifying faces, predicting behavior, flagging anomalies, and generating assessments of individuals who have not been accused of any crime, based on patterns the system has learned from data that the individuals did not consent to provide.

> *"One person. One carrier. Six months. 35,831 data points. That was 2011. The systems described in this chapter do not collect data as a byproduct. They are purpose-built to monitor entire populations."*

The Laboratory

The most extensively documented mass surveillance system in the world operates in China's Xinjiang Uyghur Autonomous Region. Since approximately 2017, the Chinese government has constructed an infrastructure of monitoring that encompasses virtually every dimension of daily life for the region's approximately 12 million Uyghur inhabitants.

The system integrates multiple surveillance technologies into a unified monitoring platform. Facial recognition cameras are installed at checkpoints, intersections, mosques, markets, residential entrances, and along roads. Mobile phones are subject to mandatory installation of monitoring software that tracks communications, contacts, and content. DNA samples, voice recordings, and iris scans are collected from the population as part of a compulsory "health check" program. Wi-Fi probes at checkpoints harvest device identifiers. Movement between cities requires passing through security stations where identity documents are scanned and biometric data is verified.

The data from these sources feeds into an integrated platform that researchers and journalists have identified by several names, including the Integrated Joint Operations Platform (IJOP). The platform applies algorithmic analysis to the collected data and generates alerts when an individual's behavior deviates from patterns the system considers normal. The deviations that trigger alerts include, according to leaked documents and research by Human Rights Watch, the Australian Strategic Policy Institute, and others: using a phone that was previously registered to another person, traveling to a neighboring country, having relatives abroad, praying more frequently than the system's baseline, receiving a phone call from a foreign number, and using a VPN.

Individuals flagged by the system have been detained in a network of facilities that the Chinese government describes as vocational education and training centers and that international observers, governments, and the UN High Commissioner for Human Rights have characterized as arbitrary detention camps. Estimates of the number of people detained range from hundreds of thousands to over one million. The detentions are based, in significant part, on algorithmic assessments generated by the surveillance platform.

The Weapon in the Pocket

Xinjiang's surveillance infrastructure is geographically bounded. A different surveillance technology operates without geographic limits, without the target's knowledge, and, in its most sophisticated form, without requiring the target to do anything at all.

Pegasus is a surveillance tool developed by the Israeli company NSO Group. It is spyware — software designed to be installed on a target's mobile phone without their knowledge or consent. In its most advanced form, Pegasus can be installed through a "zero-click" exploit — an attack that requires no action by the target. No link to click. No file to open. The phone is compromised through a vulnerability in the operating system or a messaging application, silently, without any indication to the user.

Once installed, Pegasus provides the operator with complete access to the phone's contents and capabilities: messages, emails, photographs, contacts, location data, browsing history, calendar, passwords. It can activate the phone's microphone and camera without the user's knowledge, turning the device into a real-time surveillance instrument. The phone in the target's pocket becomes, without their awareness, a microphone in the room, a camera in their hand, and a complete record of their digital life.

The Citizen Lab at the University of Toronto, Amnesty International's Security Lab, and a consortium of media organizations have documented the deployment of Pegasus against journalists, human rights activists, lawyers, political opposition figures, and heads of state across dozens of countries. Targets have included journalists investigating government corruption in Mexico, human rights defenders in Saudi

Arabia and the United Arab Emirates, opposition politicians in Rwanda and Poland, and associates of the murdered Saudi journalist Jamal Khashoggi — whose phone was reportedly compromised with Pegasus in the months before his assassination at the Saudi consulate in Istanbul in October 2018.

NSO Group maintains that Pegasus is sold exclusively to government agencies for the purpose of combating terrorism and serious crime. The documented deployments tell a different story: the technology is used, repeatedly and systematically, against the people who hold governments accountable — the journalists, the lawyers, the activists, the opposition politicians. The people who are, in the framework of this book, the keepers of democratic accountability. The watcher is deployed against the keeper.

The Inversion

Until now, the keeper has been someone who stands between a system and the people the system affects. The hardware interlock between the beam and the patient. The pilot between the autopilot and the passengers. The judge between the algorithm and the defendant. The doctor between the risk score and the patient.

Mass surveillance inverts this relationship. The system does not remove the keeper from a process. The system targets the keeper directly. Journalists who investigate state abuses are surveilled. Lawyers who defend political prisoners are surveilled. Activists who organize dissent are surveilled. The surveillance does not remove them from a loop. It undermines their capacity to function as keepers at all — by exposing their sources, mapping their networks, chilling their speech, and making them vulnerable to arrest, harassment, or worse.

This is a qualitative shift in the book's argument. The Therac-25 removed a safety mechanism. The autopilot degraded a pilot's skill. MCAS overrode the pilot's input. COMPAS deauthorized the judge. The recommendation engine outpaced the moderator. AI targeting compressed the reviewer to a rubber stamp. Each of these removed, degraded, or marginalized the keeper within a specific system.

Mass surveillance attacks the keeper *as a category*. It does not degrade the keeper's function within one system. It degrades the keeper's ability to exist as a keeper in any system — by making the act of watching, questioning, and holding power accountable visible to the power being held accountable. The keeper who knows they are watched is a keeper who self-censors, who avoids sensitive topics, who protects sources by not contacting them, who serves the function of accountability less effectively because the cost of serving it has been raised beyond what most people are willing to bear.

The Score

China's social credit system represents a different modality of surveillance — not the targeted monitoring of specific individuals but the continuous assessment of an entire population's behavior against a set of norms defined by the state.

The social credit system, as implemented across various Chinese municipalities, aggregates data from government records, financial transactions, court judgments, regulatory compliance databases, and, in some implementations, social media behavior and associations. Individuals and businesses receive scores or classifications that affect their access to services, their ability to travel, their eligibility for loans, and their public reputation. Those classified as untrustworthy face restrictions: denial of air and rail tickets, exclusion from certain schools for their children, public listing on blacklists, and reduced access to government services.

The system is not a single, unified algorithm — it is a patchwork of local implementations with varying criteria, data sources, and consequences. But the architecture is consistent: continuous data collection, algorithmic assessment of behavior against state-defined norms, and automated consequences for deviation. The human decision-maker is not in the loop for individual assessments. The system generates the classification. The consequences flow automatically.

The reader who has followed this book from the Therac-25 through COMPAS will recognize the architecture. A system that encodes the biases of the institution that built it (Invisible Bias). A system that operates faster than human review can match (Speed-Judgment Tradeoff). A system where responsibility for individual outcomes diffuses across the platform, the data sources, and the policy framework (Accountability Gap). A system whose reliability produces dependence that makes reversal difficult (Reliability Trap). A system whose assessments are based on data that is treated as ground truth without independent verification (Single Point of Trust).

All five patterns. One system. Applied to 1.4 billion people.

The Export

The surveillance infrastructure developed in Xinjiang and tested in Chinese cities has not remained within China's borders. Chinese technology companies — including Huawei, ZTE, Hikvision, and Dahua — have exported surveillance systems to governments across Africa, Southeast Asia, Central Asia, Latin America, and the Middle East. These systems include facial recognition cameras, network monitoring tools, and integrated surveillance platforms based on architectures developed for domestic Chinese use.

The Carnegie Endowment for International Peace documented in a 2019 study that AI surveillance technology had been deployed in at least 75 countries worldwide, with Chinese companies supplying technology to at least 63 of them. The deployments include countries with documented records of authoritarian governance, suppression of political opposition, and human rights abuses.

The pattern is the one this book identified in Chapter 5: the system is designed for one purpose (counterterrorism, public safety), deployed for that purpose in the initial context, and then repurposed for a different function (political surveillance, suppression of dissent) once the infrastructure is in place. The infrastructure does not distinguish between purposes. A facial recognition camera installed to identify criminal suspects works equally well to identify political protesters. A communications monitoring platform deployed to intercept terrorist messages intercepts all messages. The technology is dual-use not as an unfortunate side effect but as a structural property. The same system that protects can oppress, and the difference is determined not by the technology but by the intent of the operator — an intent that can change with a new government, a new policy, or a new definition of who constitutes a threat.

The Democratic Watcher

It would be reassuring to frame mass surveillance as an authoritarian problem — a tool of dictatorships that democracies resist. The evidence does not support this framing.

The Snowden disclosures of 2013 revealed that the US National Security Agency operated surveillance programs that collected the communications metadata of millions of American citizens without individual warrants. The UK's GCHQ operated programs that tapped undersea fiber-optic cables to intercept vast quantities of internet traffic. The "Five Eyes" intelligence alliance — the US, UK, Canada, Australia, and New Zealand — shared surveillance data across borders in ways that allowed each country to monitor activities that domestic law might prohibit it from monitoring directly.

Democratic governments deploy surveillance infrastructure for reasons that democratic publics often accept: counterterrorism, law enforcement, border security, pandemic response. The infrastructure, once deployed, persists. The legal authorities, once granted, expand. The surveillance capabilities developed for one crisis become available for the next, and the definition of what constitutes a crisis broadens over time.

The difference between surveillance in a democracy and surveillance in an authoritarian state is not the technology. It is the institutional infrastructure that allows the keeper to exist: courts that can review surveillance requests, legislatures that can limit authorities, press freedoms that protect the reporter who exposes abuses, and civil society organizations that can litigate against overreach. These are not the keepers. They are the conditions under which keeping is possible. The journalist whose phone is hacked by Pegasus is the keeper — the individual human exercising judgment, asking questions, holding a light to the system. The court is what protects her. The legislature is what constrains the watcher. The press freedom is what allows the story to be published. Surveillance tools like Pegasus are used against all of them — not because they are the system, but because they are the infrastructure that keeps the keeper alive.

The watcher watches everyone. But it watches the keeper most closely. Because the keeper is the only threat the watcher cannot predict, cannot quantify, and cannot optimize away: the human who looks at the system itself and asks whether it should exist.

> *"The difference between surveillance in a democracy and in an authoritarian state is not the technology. It is the infrastructure that allows the keeper to exist — the courts, the legislatures, the press freedoms. Pegasus targets all of them.*

The Pipeline

The previous chapter described AI systems that select targets for lethal action. This chapter describes the infrastructure that feeds those systems. The connection is not metaphorical. It is operational.

The targeting algorithms described in Chapter 9 — Gospel, Lavender, and their equivalents in other militaries — require data. The data comes from surveillance: signals intelligence, communications intercepts, location tracking, facial recognition, social network analysis. The more comprehensive the surveillance, the more data the targeting system has. The more data the targeting system has, the more targets it generates. The more targets it generates, the less time the human has to review each one.

Surveillance and targeting are not separate systems. They are stages in a pipeline. The watcher feeds the target. The target feeds the weapon. The human in the loop — the keeper who is supposed to ensure that the pipeline produces lawful, proportionate, and discriminate outcomes — sits at the end of a process that has already determined the conclusion. The keeper's twenty-second review is the final stage of a pipeline that began with a camera on a street corner, a phone in a pocket, or a metadata record on a server.

This is why Anthropic's two red lines — no autonomous weapons and no mass surveillance — are not two separate positions. They are one position. The surveillance and the targeting are a single system with a single logic: identify the target and act on the identification. Drawing a line against one without drawing a line against the other is drawing no line at all.

From the Watcher to the Light

This chapter and the one before it have described the most consequential applications of automated systems: the selection of targets for killing and the monitoring of populations for control. In both cases, the keeper was not merely removed but targeted — the surveillance infrastructure described in this chapter is designed, in part, to identify and neutralize the people who might hold the system accountable.

The book has now traced the pattern through every domain it set out to examine: the systems that move us, the systems that heal us, the systems that judge us, the systems that watch us, and the systems that fight for us. In each domain, the keeper was removed, degraded, or overwhelmed. In each case, the cost was borne by people who did not consent to the removal and often did not know it had occurred.

The final chapter returns to where this book began: a lighthouse on a rock in the Atlantic, and a man who kept the light burning because he understood that the light was not for him. The question is whether, after everything this book has documented, anyone will keep the light.

The Last Keeper

The Light

The closing argument. The return to the lighthouse.
The five patterns restated through the full arc.
And the question this trilogy exists to ask.

"The light was not for him. The light was for the ships."
— *Chapter 1*

The Light

The lighthouse at Smalls still stands. It is no longer the wooden structure on stilts that Henry Whiteside built in 1776, the one Thomas Howell clung to through Atlantic storms while keeping the light burning beside his dead companion. That lighthouse was replaced in 1861 by a stone tower, and the stone tower was automated in 1987. The last keepers left. The light now operates by electricity, monitored remotely from the mainland. It is more reliable than any human keeper could have maintained it. It does not get tired. It does not get sick. It does not need food or rest or relief.

The light shines every night. On average, it is better than the keeper it replaced.

This book has not been about the average.

> *"The light shines every night. On average, it is better than the keeper it replaced.*
> *This book has not been about the average."*

What the Book Found

This book began with a question: what happens when the person in the loop is removed? It traced that question through twelve domains, across more than a century, from a lighthouse on a rock to an AI system that generates military targets. What it found was not a collection of unrelated failures. It found a pattern — five patterns, in fact — that repeat with a consistency that would be remarkable if it were not also predictable.

The **Reliability Trap** appeared on the Titanic, where the ship's engineering record produced the confidence that reduced the lifeboats. It appeared in the cockpit of Air France 447, where the autopilot's reliability eroded the pilots' proficiency. It appeared in the courtroom, where the algorithm's consistency displaced the judge's willingness to exercise independent judgment. It appeared in military targeting, where the system's efficiency compressed human review to twenty seconds. The trap is the same in every domain: the system's success is the mechanism of the human's decline.

The **Single Point of Trust** appeared in the Therac-25, where software replaced hardware interlocks. It appeared in the 737 MAX, where one sensor controlled a system that could overpower the pilots. It appeared in AI targeting, where a single algorithmic assessment is treated as equivalent to verified intelligence. The pattern is the same: trust concentrated in a single system, without independent verification, is trust that will eventually be betrayed.

The **Invisible Bias** appeared in COMPAS, where the algorithm laundered racial discrimination through proxy variables. It appeared in the healthcare algorithm, where cost was confused with need. It appeared in facial recognition, where training data composed predominantly of lighter-skinned faces produced a system that was selectively blind. The pattern: the data reflects the world as it is, the algorithm learns the world as it is, and the output reproduces the world's inequities while appearing to transcend them.

The **Speed-Judgment Tradeoff** appeared in the Therac-25, where the operator had seconds. It appeared in Air France 447, where the pilots had three minutes. It appeared on Facebook, where the algorithm distributed content in milliseconds. It appeared in AI targeting, where the system generates targets faster than humans can evaluate them. The trajectory is clear: the window for human judgment is shrinking, and at sufficient speed, the human is not in the loop but beside it.

This pattern poses a challenge this book must confront directly. If the system operates faster than human thought, what does "keeping the keeper" mean? The answer is not faster humans. The answer is systems that are not permitted to execute consequential decisions at speeds that preclude human review. The objection is obvious: the adversary who does not slow down gains the advantage. The response is equally uncomfortable: the advantage of speed without judgment is the advantage of a weapon without a safety. It will fire faster. It will also fire at the wrong target faster. The keeper's function is not to match the machine's speed. The keeper's function is to ensure that the machine does not exceed the speed of judgment.

The **Accountability Gap** appeared in the Therac-25, where the manufacturer could not reproduce the error and declared the machine safe. It appeared in the 737 MAX, where responsibility was distributed across Boeing, the FAA, the airline, and the sensor manufacturer. It appeared in the Myanmar genocide, where Facebook's algorithmic amplification introduced a new kind of causation for which existing frameworks have no vocabulary. It appeared in military targeting, where the chain of responsibility dissolves across the system. The pattern: the more automated the system, the more diffuse the accountability, and the less likely that anyone bears the cost of the failure.

The Arc

The book also traced an arc — a progression in the relationship between humans and the systems they build.

In Part I, the keeper was **present but overruled**. The Titanic's officers had the warnings. They trusted the ship. The keeper's judgment was overridden by institutional confidence in engineering.

In Part II, the keeper was **removed, degraded, and overridden** in sequence. The Therac-25 removed the hardware safety layer. Air France 447 degraded the pilots' manual skills. The 737 MAX overpowered the pilots' physical inputs. Each step left the human with less capacity to intervene.

In Part III, the keeper was **deauthorized and outpaced**. The judge was given a number that discouraged independent judgment. The doctor was given a metric that displaced clinical intuition. The content moderator was outrun by an algorithm that distributed content faster than any human could review. The system did not remove the keeper. It made the keeper's role nominal.

In Part IV, the keeper was **rubber-stamped, targeted, and overwhelmed**. The military reviewer had twenty seconds per target. The surveillance system identified and undermined the people whose function is to hold systems accountable. And the complexity of modern software crossed a threshold beyond which human verification became not difficult but impossible.

The keeper has functioned, in the cases this book has documented, only when an individual with deep expertise and institutional independence chose to look beneath the number. The question is not whether such keepers exist. They do. The question is whether the systems being built will make room for them — or whether the algorithm will make them unnecessary before the next one arrives.

The arc moves in one direction: from the keeper present and overruled, to the keeper absent and irrelevant. Every era has justified the movement in the same terms: the system is faster, cheaper, more reliable, more consistent, more objective. The justification is, on the merits, largely correct. The machine *is* better than the human on average, in most conditions, most of the time. The book does not dispute this.

The book disputes the conclusion drawn from it.

The Conclusion That Does Not Follow

The conclusion drawn, in every era and every domain this book has examined, is that because the machine is better on average, the human backup can be reduced. The lifeboats can be cut from 48 to 20. The hardware interlock can be replaced with software. The pilot can be allowed to deskill. The judge can defer to the score. The reviewer can approve in twenty seconds. The content can be distributed without human review. The surveillance can operate without individual warrants.

This conclusion does not follow.

The value of the keeper is not measured by average performance. It is measured by performance at the margin — at the moment when the system encounters conditions it was not designed for, when the data is wrong, when the assumption fails, when the metric is a proxy for the wrong thing. The keeper's value is concentrated in the tail of the distribution: the 0.01% of events where the system is not merely imperfect but catastrophically wrong. And because the cost of catastrophic failure — in lives, in liberty, in justice — is orders of magnitude larger than the cost of maintaining the keeper, the economic calculation that justifies removing the keeper is a calculation that discounts the catastrophe.

It discounts the catastrophe because the catastrophe is rare. It is rare because the system is reliable. The system is reliable because it was well-designed. And the conclusion is: because the system is well-designed, the keeper is unnecessary. But the keeper is what makes the system survivable when the design fails. And the design always, eventually, fails.

Not on most nights. Not on the average night. On one night. The night the ice crystals block the pitot tubes. The night the sensor gives the wrong reading. The night the code encounters a value it was not written for. The night the algorithm confuses cost with need. The night the engagement metric amplifies a call for genocide. The night the confidence score is wrong and the strike kills a family.

On that night, the keeper is the only thing standing between the system's failure and its consequence. And in every case the evidence shows: on that night, the keeper was not there.

What the Keeper Must Become

This book does not argue against automation. The autopilot saves more lives than it costs. The algorithm processes more data than any human can. The surveillance system identifies genuine threats that human intelligence misses. The argument is not that machines should be removed. The argument is that the keeper must remain.

But the keeper of the twenty-first century cannot be the keeper of the nineteenth. Thomas Howell kept one light on one rock. The these systems serve billions of people simultaneously, operate at speeds no human can match, and contain complexities no human can fully understand. The lighthouse keeper model — one person, one system, continuous watchfulness — does not scale.

The keeper must evolve. The keeper must become an institution, not just an individual. It must become a set of structures: independent audit bodies with the authority and technical capacity to examine algorithmic systems before and after deployment. Mandatory disclosure requirements that make the behavior of consequential algorithms visible to the people they affect. Red teams empowered to find the failure before the failure finds the patient, the passenger, the defendant, the civilian. Kill switches that allow human override of automated systems in safety-critical domains, tested not just for existence but for physical executability — because a cutout switch that cannot be turned against the aerodynamic forces the system has created is not a safety mechanism but a fiction.

The keeper must also become a culture — an institutional commitment to the principle that the efficiency of automation does not justify the elimination of human judgment in domains where the cost of failure is measured in lives. Aviation learned this, imperfectly and at great cost. Medicine is learning it. Criminal justice is resisting it. Military operations are moving in the opposite direction. And the technology industry, which builds the systems that all other domains depend on, has not yet decided.

The Company That Drew a Line

In February 2026, a company that builds AI systems was offered a contract to provide its technology for military applications. The company's leadership identified two specific uses that were incompatible with their assessment of where the human must remain in the loop: autonomous weapons targeting and mass surveillance of civilian populations. They said no to those two uses.

The response from the defense establishment was pressure to designate the company as a supply-chain risk. The response from the public was different. Millions of people chose the company's product specifically because it had drawn a line. The market rewarded the keeper.

This is a data point, not a conclusion. One company. One decision. One moment in which the economic incentive and the ethical position aligned. The documented history provides no basis for confidence that this alignment will hold. The economic pressures to remove the keeper are persistent and structural. The political pressures are real. The institutional incentives favor efficiency over caution, speed over judgment, automation over oversight.

But it happened. A company that builds the technology said: not this use. Not this loop. The human stays. And the public, in numbers large enough to measure, said: yes. That is the company we trust. That is the product we choose.

The keeper's survival may depend less on regulation than on this: on the discovery that people, given the choice, prefer the system that keeps the human in the loop to the system that removes them. That the market for trust is larger than the market for efficiency. That the keeper is not a cost but a value — a thing people will pay for, choose, and defend.

Whether this is enough remains to be seen.

The Deeper Question

This book has documented the cost of removing the keeper. But it has not answered the question that the evidence inevitably raises: *why* does the removal keep happening?

The keeper was removed from the Therac-25 because the software was believed to be sufficient. The keeper was degraded in the cockpit because the autopilot was more reliable. The keeper was overridden in the 737 MAX because the schedule held. The keeper was deauthorized in the courtroom because the algorithm was more consistent. The keeper was outpaced on the platform because the engagement metric moved faster than any human could review. In every case, the reasoning was the same: the machine is better on average.

But behind that reasoning is a deeper pattern — a pattern that is not about individual machines or individual decisions but about the concentration of computational power itself. Every era of computing has produced a cycle: the capability distributes, and then the control reconcentrates. The mainframe concentrated power in a handful of companies. The personal computer appeared to distribute it. The internet appeared to distribute it further. And then five companies recaptured the infrastructure, the data, and the distribution channels. The keeper is removed not by accident but by the structural economics of each era's reconcentration — because the entities that accumulate power are the entities that benefit most from replacing human judgment with machine efficiency.

This book documented the cost. The deeper question — the five-thousand-year history of how computational power concentrates, why the pattern repeats, and whether the next era of computing will make the concentration permanent — is a question that demands its own investigation. It is the question that haunts the evidence this book has presented: not just what happens when the keeper is removed, but who benefits from the removal, and whether the accumulation of power that drives the removal can be governed before it becomes irreversible.

The Light

Thomas Howell kept the light burning through storms that would have killed a lesser person, beside a body that was decomposing in the salt air, on a rock twenty miles from shore with no relief and no communication with the mainland. He did this because the light was not for him. The light was for the ships.

The ships could not see the rocks. The ships could not know the currents. The ships depended on a human being who sat in the dark, in the cold, in the storm, and kept the light burning because someone had to. Not because any analysis had determined that the keeper's salary was justified by the expected reduction in shipwrecks. Because without the light, the ships would break on the rocks. And the keeper was the light.

The light at Smalls still shines. The mechanism is reliable. The keeper is gone. And this book has spent twelve chapters documenting what happens, in domain after domain, era after era, when the keeper is gone and the mechanism encounters a night it was not designed for.

The nights will come. They always do. Not one night — that framing is too comfortable, too contained, too much like a story with an ending. The failures will keep arriving for as long as the systems operate, because no system is perfect and the world is not static. The ice crystals will form in the pitot tubes. The sensor will give the wrong reading. The algorithm will confuse cost with need. The engagement metric will amplify the call for violence. The targeting system will assign a high confidence score to the wrong person. The surveillance system will flag the journalist, the lawyer, the activist. The code will encounter a value it was not written for.

These are not hypothetical failures. Every one of them has already happened. Every one of them has already killed, harmed, imprisoned, or surveilled people who did not consent to the system that failed them. The question is not whether the failures will continue. The question is whether, when those nights come — and they will keep coming — there will be a keeper watching the light.

"The nights will come. They always do. The failures will keep arriving for as long as the systems operate. When those nights come — and they will keep coming — will there be a keeper watching the light?"

What to Read Next

This book asked what happens when the human is removed from the loop. If the evidence it presented left you with the question it left me with — *why does the removal keep happening?* — then the answer may lie in a history far older and far larger than the century of failures documented here.

Who Holds the Jar? traces the concentration of computational power across five thousand years — from the shepherd's clay counting jar in ancient Mesopotamia through every era of computing to the quantum and biological substrates that may define the next. It identifies a pattern that has repeated in every era: power distributes, then reconcentrates. Each cycle appears to democratize. Each cycle produces reconcentration. And the current cycle — AI, quantum computing, biological computation — may produce a concentration of power that, for the first time in history, cannot be reversed.

If *The Last Keeper* is the human cost, *Who Holds the Jar?* is the reason the cost keeps being paid. It is available now as the second book in *The Cost of the Machine* series.

The Cost of the Machine

Book One: The Last Keeper

Book Two: Who Holds the Jar?

Book Three: The Great Convergence

Pumulo Sikaneta

References & Source Attribution

About the Author

Pumulo Sikaneta works in technology strategy, where he has spent years studying how organizations adopt, govern, and are transformed by the systems they build. His professional work in partner enablement and solutions architecture has given him a close-up view of the gap between what technology promises and what it delivers — and of the humans who stand in that gap.

The Last Keeper is the first book in *The Cost of the Machine*, a series that examines the human consequences of automated decision-making. It began as a question — why does the same pattern of failure repeat across every domain where machines replace human judgment? — and became a book when the evidence demanded more than a question.

The author's brother, Dr. Tabo Sikaneta, is a nephrologist at the Scarborough Health Network in Toronto and a founding member of the African Caribbean Kidney Association. His two decades of clinical work documenting the disproportionate burden of kidney disease in immigrant and non-White communities — including his 2025 *BMJ Open* study demonstrating that these disparities persist even within Canada's universal healthcare system — provided the human evidence that grounds Chapter 7's argument: that no algorithm can be more just than the society whose data it consumes.

The author was born in Zambia, raised in Canada, and lives in the United States. He believes that the most important skill in any field is the willingness to be corrected.

Book Two

Who Holds the Jar?

The Machine Evolves

Who Holds the Jar?

Who Holds the Jar?

The Machine Evolves

A Natural History of Computation — and a Warning

Pumulo Sikaneta

*For those who count the uncounted —
and ask who holds the jar when the counting
determines what counts.*

Contents

The Machine Evolves

Author's Preface

Before We Begin

This book builds a single argument across fourteen chapters, and it rewards sequential reading. If you begin in the middle, you will encounter ideas that depend on foundations laid in earlier chapters — not because the writing is inaccessible, but because the argument is cumulative. Each chapter adds a layer. The layers support each other. I recommend starting at the beginning.

The argument, in brief, is this: the evolution of computing follows repeating patterns that can be identified, tracked, and used to anticipate what is coming. The most consequential of these patterns involves power — who holds computational capability, how it distributes, and how it reconcentrates. I trace this pattern from the oldest known mathematical artifacts, roughly twenty thousand years old, through mechanical calculators, electronic computers, the internet, artificial intelligence, and into the emerging substrates of quantum and biological computing. The trajectory leads to a warning that I believe is both original and urgent: the greatest danger of advanced computation is not that a machine becomes conscious. It is that a small number of humans gain access to computational power so asymmetric that it constitutes a new kind of dominance — invisible, unverifiable, and potentially permanent.

Who This Book Is For

I wrote this book for the person who uses a computer every day and has never once thought about why it works the way it does. For the executive who makes decisions about technology without understanding the physical forces beneath it. For the curious reader who follows the news about AI, quantum computing, and the technology industry and wants a framework that makes sense of it all. And for the technologist who knows the parts but has never seen the whole arc — the five-thousand-year trajectory from counting pebbles to manipulating atoms.

I have made every effort to explain technical concepts in plain language, using familiar objects and human stories as entry points. You do not need a background in computer science, physics, or engineering to read this book. You need only curiosity and the willingness to follow an argument across fourteen chapters. If a concept requires jargon, I define it immediately. If an idea requires an analogy, I provide one and then deepen it. I do not condescend. The reader is intelligent. They simply have not been told this story before.

What This Book Is Not

This is not a textbook, a technical manual, or a comprehensive encyclopedia of computing history. Excellent books in each of those categories already exist, and I acknowledge them in the References document that accompanies this series. In particular, Walter Isaacson's *The Innovators* (2014), Martin Campbell-Kelly and William Aspray's *Computer* (4th edition, 2023), and James Gleick's *The Information* (2011) cover much of the same historical territory with different emphases and different theses. I have learned from all of them. Where our factual coverage overlaps — as it must, since there is only one history of the transistor and only one first internet message — the narrative voice, the analytical framework, and the forward-looking argument are my own.

This book's contribution is not the facts but the framework: the five repeating patterns, the distribute-then-recapture power dynamic, the convergence spiral from nature through engineering back to nature, and the computational omniscience warning of the final chapters. These, to my knowledge, are original to this work. I welcome correction.

A Note on Methodology

The method of this book owes a significant debt to Ray Dalio, whose work on identifying repeating patterns in historical data and using them to generate testable predictions shaped my approach. It owes an equal debt to Yuval Noah Harari, whose *Sapiens* demonstrated that deep history could be told with a provocative thesis, accessible prose, and genuine intellectual ambition. I have tried to honor both influences: Dalio's rigor in pattern identification and prediction, Harari's accessibility and narrative breadth. The synthesis and the conclusions are my own, and so are the errors.

I present the frameworks in this book not as certainties but as the best instruments I can construct from the evidence available. They are meant to be tested, challenged, and revised as events unfold. I would rather be wrong about the specific predictions and useful in providing the framework than right about the predictions and useless in the face of them.

The tagline I use in all my published work is: *Curious by nature. Correctable by choice. Always learning. Often wrong. Let's discuss.* I mean every word of it. This book represents my current best understanding of a five-thousand-year trajectory, and I hold it loosely. The patterns I have identified may prove to be less universal than I believe. The predictions may prove to be wrong. The warning may prove to be premature or, conversely, too late.

What I am confident of is the question. Whether or not I have the right answer, I believe *who holds the jar?* is the right question for this moment in the history of computation. If this book does nothing more than cause you to ask that question — and to keep asking it as the technology evolves — it will have accomplished what I set out to do.

March 2026

Pumulo Sikaneta

BEFORE ELECTRICITY

The Machine Evolves

The Pebble and the Problem

On a diver who surfaced screaming about corpses, a two-thousand-year-old
computer pulled from the sea, a shepherd who needed to know if he'd
lost a sheep, and why the most powerful machine in history began
with a handful of clay tokens in a jar

"The purpose of computing is insight, not numbers."
— Richard Hamming

The Pebble and the Problem

On a diver who surfaced screaming about corpses, and why the most powerful machine in history began with a handful of clay tokens in a jar

Around Easter of 1900, a storm blew a small boat full of sponge divers off course in the Aegean. Captain Dimitrios Kontos, from the island of Symi, had been heading for the sponge beds off the North African coast. Instead he found himself anchored in the lee of a rocky island called Antikythera, between Crete and the Greek mainland, waiting for the wind to drop. Kontos was not the kind of man who wasted time. While the storm held, his crew would dive.

Elias Stadiatis pulled on the newfangled diving suit — a canvas shell with a copper helmet and a rubber hose feeding air from a hand pump on deck — and dropped into forty-five metres of dark water. He was down for less than a minute. When they hauled him to the surface, he was frantic. He described a scene on the seafloor that made no sense: heaps of rotting corpses and dead horses, scattered among the rocks, staring up at him.

The crew thought he was delirious. Nitrogen narcosis at that depth could make a diver see things. Kontos did what any captain would: he suited up and went down himself. He stayed longer than Stadiatis had. When they pulled him back aboard, he was holding the arm of a bronze statue.

The corpses were not corpses. The horses were not horses. They were sculptures — marble and bronze figures from a Roman cargo ship that had gone down more than two thousand years earlier, their surfaces encrusted with centuries of marine growth, their features blurred into something that looked, in the dim green light of forty-five metres, exactly like the dead.

Kontos kept the wreck's location secret. He sailed to Athens, showed the bronze arm to the authorities, and negotiated: his divers would do the recovery work. Greece was in financial trouble and political turmoil; a haul of ancient treasures — reminders of past glories — was exactly what the government needed. The Hellenic Navy dispatched ships. Recovery began in November 1900 and stretched into the following year. The work was brutal. One diver, Georgios Kritikos, died from decompression sickness. Two others were paralyzed. But the team brought up dozens of bronze and marble statues, pottery, jewellery, glassware, and coins from the first century BCE.

Among the treasures was a lump of corroded bronze about the size of a shoebox. It looked like nothing. Nobody paid it any attention. It was cataloged, stored, and forgotten.

It was the most sophisticated piece of technology built in the ancient world.

The Machine at the Bottom of the Sea

In May 1902, a Greek archaeologist named Valerios Stais was examining the Antikythera haul in a workroom at the National Archaeological Museum in Athens when he noticed something odd about the corroded lump. As the wood casing had dried in the open air, it had cracked apart, revealing bronze fragments with tiny inscriptions and what appeared to be the precision gears of a mechanical clock.

It took the better part of a century to understand what Stais had found. In the 1950s, a British historian of science named Derek de Solla Price became obsessed with the fragments. Working with a Greek radiographer, he took the first X-rays of the device. In 1974 he published his findings: the corroded lump was a mechanical computer, built around 100 BCE, containing at least thirty interlocking bronze gears, precisely machined, that could predict the positions of the sun and moon, the timing of eclipses, and the schedule of the ancient Olympic Games.

A person in the first century BCE — a world with no electricity, no printing press, no telescope — could turn a hand crank and know, with considerable accuracy, where the sun and moon would be on any given date in the future. The machine stored astronomical data in the arrangement of its gears. It retrieved the data through the position of its dials. It compared celestial cycles through the ratios of its gear trains. And it transformed inputs — turning the crank — into outputs: positions on its display. Once the gears were set, the machine ran on its own logic. The human turned the crank. The mechanism did the rest.

Nothing of comparable complexity would appear in the historical record for over a thousand years.

Most histories of computing are where the narrative moves on. The Antikythera mechanism teaches because the Antikythera mechanism teaches two lessons that will echo through every chapter of this book.

What the Mechanism Teaches

The first lesson is about power.

In the ancient world, predicting celestial events was not an academic exercise. It was political and religious authority. A priest who could announce, weeks in advance, that the moon would vanish from the sky held enormous influence over a population that understood eclipses as omens from the gods. An astrologer who could tell a king the most auspicious day to launch a military campaign wielded real power — not because the stars actually governed outcomes, but because the king believed they did, and acted accordingly.

The Antikythera mechanism was, among other things, a power tool. Whoever operated it could predict what others could not. And the gap between what you can predict and what the people around you can predict is the oldest form of strategic advantage. It is the same advantage held by every institution, government, and corporation that has controlled computation in the centuries since.

The second lesson is about loss.

The Antikythera mechanism was built in a world that, within a few centuries, would lose the ability to build it. The precision gear-cutting techniques, the astronomical knowledge encoded in the gear ratios, the manufacturing skill required to mesh thirty gears in a casing the size of a shoebox — all of this knowledge vanished. It did not fade gradually. It was lost in the collapse of the Greco-Roman world and the fragmentation of the networks that had sustained it. When the mechanism was pulled from the sea in 1901, no scholar alive could explain how it worked. It took another hundred years of research to understand what a craftsman in 100 BCE had built with hand tools.

Technological progress is not a ratchet. It does not only go forward. It can be lost — through war, through civilizational collapse, through the simple failure to transmit knowledge from one generation to the next. The Antikythera mechanism is a reminder that what we have built is not guaranteed to survive. That is a truth worth holding as we trace the story forward through substrates of increasing power and fragility.

The Shepherd and the Jar

But the Antikythera mechanism, extraordinary as it was, is not where this story begins. It begins much earlier, much simpler, and much closer to the ground.

Somewhere around five thousand years ago, in what is now southern Iraq, a shepherd faced a problem that his brain could not reliably solve.

He had a flock. The flock left in the morning. The flock returned in the evening. And every evening, he needed to answer a simple question: did they all come back?

The human mind is a remarkable instrument. It can recognize a face it has not seen in twenty years, compose a song from nothing, and sense danger from the faintest change in another person's expression. What it cannot do, with any reliability, is count to sixty-three and remember the result twelve hours later. Not every day. Not without error. Not when the sheep look alike and you are tired and it is getting dark.

The shepherd's solution was elegant. Each morning, as the sheep filed out through the gate, he dropped a small clay token into a jar — one token for each animal. Each evening, as they returned, he removed a token for each one that came back. If the jar was empty at the end, the flock was complete. If tokens remained, sheep were missing. The shepherd did not need to count. He did not need to remember a number. He needed to compare two quantities, and the jar did it for him.

That jar of clay tokens is a computer.

Not metaphorically. Functionally. It stores information — the count of the flock. It has an input mechanism — dropping a token in. It has an output mechanism — the tokens that remain. And it has a purpose: answering a question that the human mind could not reliably answer on its own. Every machine we have built since — from the abacus to the smartphone in your pocket to a quantum processor manipulating individual atoms — is an elaboration of that jar.

Begin with a shepherd and a handful of pebbles, because I believe the history of computing is almost always told from the wrong starting point. Most accounts begin with Charles Babbage in the 1830s, or with the electronic computers of the 1940s, as though computation were an invention of the industrial age. It is not. Computation is one of the oldest human activities — older than writing, older than agriculture, possibly older than spoken language as we know it. And the reason it is so old is that it solves the most fundamental problem any organism faces: the gap between what the mind can imagine and what the mind can reliably execute.

That gap is where every computer lives. It is where every computer has always lived. And the story of how we have tried to close it — with pebbles, with gears, with vacuum tubes, with silicon, and soon with the laws of quantum physics and the molecules of life itself — is, I believe, the most consequential story in human history.

Older Than Writing

The oldest mathematical artifact ever found is a small, dark bone roughly the length of a pencil, discovered in 1960 in what is now the Democratic Republic of Congo. It is called the Ishango bone, and it is approximately twenty thousand years old.

Carved along its surface are three columns of notches — grouped in patterns that have fascinated and divided mathematicians for decades. One column contains groups of 11, 13, 17, and 19 — the prime numbers between 10 and 20. Another contains groups that could represent a doubling sequence. Whether this is intentional mathematics or coincidental grouping is still debated. What is not debated is that someone, twenty thousand years ago, was scratching a systematic record of quantities onto a bone. They were storing information outside their own mind. They were computing.

Twenty thousand years ago, there were no cities, no agriculture, no written language, no metal tools. Humans lived in small groups, hunted with stone-tipped spears, and wore animal skins. And somebody, in that world, felt the need to record numerical information in a durable medium. The urge to compute is not a product of civilization. It is a precondition for it.

The Mesopotamian shepherds came much later — around 3500 to 3000 BCE — and their counting tokens grew into something more sophisticated. Archaeologists have found clay tokens of different shapes: spheres for grain measures, cones for small amounts of oil, discs for livestock. The tokens were placed inside hollow clay balls called *bullae*, which were then sealed. To know what was inside without breaking the seal, accountants began pressing the tokens into the surface of the wet clay before enclosing them — creating a visible record on the outside of what was stored on the inside.

Some scholars believe this is how writing was born. The impressions on the outside of the *bullae* became standardized. The tokens inside became unnecessary. The marks alone carried the meaning. If this theory is correct — and the evidence is compelling — then writing itself is a byproduct of computation. Humans did not first learn to write and then use writing for accounting. They first learned to account, and the marks they made became writing.

Which means that the most transformative information technology in human history — the written word — may be a side effect of humanity's need to count.

The Reckoning Frame

The abacus appeared in Sumeria sometime between 2700 and 2300 BCE, and it represents a genuine leap. The counting tokens and tally marks that preceded it could store and compare, but they were slow and clumsy at transformation. To add two groups of tokens, you physically combined them. To subtract, you physically removed. For anything more complex, the system collapsed under its own weight.

The abacus changed this by introducing a crucial idea: positional representation. A bead's value depended on which column it occupied. A bead in the rightmost column meant one. The same bead, moved to the next column, meant ten. The next column, a hundred. This is the same idea behind the numbers you use every day — the digit 3 means three in the ones place, thirty in the tens place, three hundred in the hundreds place. The abacus was the first machine to exploit this principle.

How effective was it? In 1946, in a widely publicized contest reported across American and Japanese media, a Japanese postal clerk named Kiyoshi Matsuzaki sat across a table from an American soldier operating an electric calculating machine — the best electronic arithmetic technology the United States military had. They competed across five categories of calculation: addition, subtraction, multiplication, division, and a mixed category combining all four. Matsuzaki, using an abacus, won four of the five. The ancient tool, in trained hands, was still faster than twentieth-century electronics.

The abacus also appeared independently in China, in Rome, in Mesoamerica, and in several other civilizations that had no contact with each other. This independent invention is significant. It means the abacus is not a cultural artifact of a particular society. It is a convergent solution — the way that wings evolved independently in birds, bats, and insects because flight solves a problem that many different organisms face. The abacus is what computation looks like when human hands are the only available moving parts.

But there is one critical thing the abacus cannot do. It cannot operate without a human. Every bead that moves, moves because a finger pushes it. The machine has no autonomy. It performs no operation on its own. It is, in the most precise sense, a tool — an extension of the human hand and mind, incapable of independent action.

The first evidence that a machine could act on its own came from the wreck at the bottom of the Aegean that opened this chapter. The Antikythera mechanism, once its gears were set, ran on its own logic. Turn the crank and the output followed from the input with no further human judgment required. That is automation. And the gap between a tool that requires a human hand on every bead and a mechanism that runs on its own logic is the gap that every subsequent chapter of this book will explore — because it is the same gap that exists today between a spreadsheet and an AI system that makes decisions without being asked.

The Substrate Principle

This chapter has described four different computing technologies: clay tokens, tally marks on bone, the abacus, and the Antikythera mechanism. They span roughly twenty thousand years. They were built by cultures that had no knowledge of each other. And yet they are all doing the same thing — storing, retrieving, comparing, and transforming information, performed on different physical materials.

The physical material matters. Clay tokens can store and compare but cannot transform quickly. The abacus can transform rapidly but cannot operate without a human hand. The Antikythera mechanism can operate automatically but is limited to a single domain — astronomy — hardwired into its gears. Each material enables certain capabilities and constrains others. The material is what I will call, throughout this book, the *substrate* — the physical medium on which computation runs.

Here is the pattern that the rest of this book will trace: every substrate eventually reaches a ceiling. A physical limit, imposed not by human imagination but by the laws of nature, beyond which the material cannot be pushed. Clay tokens cannot perform multiplication. An abacus cannot operate autonomously. Gears cannot compute faster than the speed of their mechanical action. When the ceiling is reached, progress does not stop. The substrate changes. And when the substrate changes, everything changes — not just how fast we compute, but who can compute, what gets computed, and who controls the results.

The change is not gradual. Not the same machine getting better. It is a phase transition — like water becoming steam. The old substrate does not get incrementally better. It is replaced by something operating on fundamentally different physics. And each time this happens, the consequences ripple far beyond the machine itself, into economics, politics, warfare, and the basic structure of daily life.

We are approaching one of these transitions right now. Actually, we are approaching two simultaneously — quantum computing and biological computing — which is unprecedented in the history of computation. But before we can understand what is coming, we need to understand what came before, and why each transition followed the pattern it did.

Who Holds the Jar

There is one more pattern, and it is the one that gives this book its title.

The shepherd's jar was his. He made it, he filled it, he read the result. The computation served the person who performed it. But the Antikythera mechanism was not a shepherd's tool. It was expensive, complex, and rare. Whoever owned it had access to predictions that nobody else could make. The mechanism's power was not merely technical. It was political — the power to see what others could not, to anticipate what others could not, to act on knowledge that others did not have.

Between the shepherd's jar and the Antikythera mechanism, something shifted. Computation went from personal to institutional. From the hand of the person with the problem to the hand of the person with the resources to build the machine. This shift — the movement of computational power from the many to the few, and occasionally back again, in a cycle that repeats in every era this book will document — is the central thread of the story that follows.

Call it the *distribute-then-recapture* cycle. A new substrate arrives. It distributes computational capability more widely than before — to more people, more cheaply, more accessibly. And then, over time, the control of that capability reconcentrates in the hands of a few. The tokens changed. The jar changed. The question did not: who holds it?

> *"A new substrate arrives. It distributes computational capability more widely than before. And then, over time, the control reconcentrates in the hands of a few. The tokens changed. The jar changed. The question did not."*

Elias Stadiatis surfaced screaming about corpses and died of no fame anyone recorded. Georgios Kritikos died of decompression sickness recovering the treasures. The lump of corroded bronze sat in a museum storeroom for two years before anyone noticed the gears inside it, and for another seventy before anyone understood what they meant. The most sophisticated computer of the ancient world was invisible to the people who found it. They were looking for statues.

This is another pattern worth remembering. The most consequential technology in any era is often the one that nobody recognizes at first — because it does not look like what people expect power to look like. It looks like a corroded lump. It looks like a jar of pebbles. It looks like a toy. Until it isn't.

The next chapter moves forward three thousand years, to a different kind of computing, a different kind of builder, and a different kind of ambition. A nineteen-year-old in France builds a machine to save his father from drowning in arithmetic. A philosopher in Germany invents a number system the world will ignore for 250 years. And a woman looks at a machine made of brass and sees something no one else can see: a future in which machines would not merely calculate, but create.

The Machine Evolves

The Clock and the Calculation

On a nineteen-year-old who built a calculator to save his father
from drowning in arithmetic, a philosopher who invented a number
system the world would ignore for 250 years, and a woman who
looked at a machine made of brass and saw the future

*"At each increase of knowledge, as well as on the contrivance
of every new tool, human labor becomes abridged."*
— *Charles Babbage, On the Economy of Machinery and Manufactures, 1832*

The Clock and the Calculation

On a nineteen-year-old who built a calculator to save his father from drowning in arithmetic

In 1639, in the French city of Rouen, a sixteen-year-old named Blaise Pascal watched his father die by arithmetic.

Not literally. Not yet. But the trajectory was visible to anyone who looked. Étienne Pascal had recently been appointed the king's tax commissioner for upper Normandy — a prestigious position that came with a soul-crushing obligation. The work required endless columns of addition and subtraction, tallying tax receipts across hundreds of parishes, checking figures that had been checked before, searching for errors that crept in every time a human hand copied a number from one ledger to another. The work was important. It was also grinding the man down. Not dramatically, not the way wars or plagues kill. Slowly. Through the monotony of arithmetic that the human brain can perform but was never designed to sustain.

The boy watched. And then the boy decided to build a machine that could do it instead.

He was nineteen when the first working model was complete. He called it the Pascaline — a brass box, roughly the size of a large book, with a row of numbered wheels on top. You dialled in a number by turning the wheels with a stylus, the way you might dial a rotary telephone. Internal gears carried the tens — when a wheel turned past nine, it advanced the wheel to its left by one, exactly the way an odometer in a car rolls over from 999 to 1000. The machine could add and, through a method of complements that Pascal devised, subtract. It was not fast. It was not always reliable. The gears sometimes jammed. But it worked.

Pascal would build roughly fifty Pascalines over the next decade, each a small refinement of the last. Only about twenty were completed and fewer still were sold. The machines were expensive — far more costly than hiring a clerk to do the same arithmetic — and clerks, unlike brass gears, could also answer questions, carry messages, and make coffee. The Pascaline was a commercial failure.

It was also one of the most important machines ever built. Not for what it did — adding and subtracting are modest accomplishments — but for what it proved. A mechanical device, built from gears and springs, could perform an operation that had previously required a human mind. The operation was simple. The principle was not. The Pascaline demonstrated that arithmetic was not an act of thought. It was a physical process — a sequence of mechanical steps that could be performed by anything capable of executing those steps in the correct order. The mind was not necessary. The mechanism was sufficient.

That insight — that a mental operation can be reduced to a mechanical one — is the foundation of every computer that has ever existed. Pascal did not articulate it in those terms. He was trying to save his father. But the principle he demonstrated would take three hundred years to reach its full expression, and when it did, it would change everything.

The Philosopher and the Coin

Fifty years after Pascal, a German philosopher and mathematician named Gottfried Wilhelm Leibniz built a more ambitious machine. His Step Reckoner, completed in prototype around 1694, could not only add and subtract but also multiply and divide. It used a clever mechanism called a stepped drum — a cylinder with teeth of varying lengths — that could engage different numbers of gear teeth depending on how far it was rotated. The machine was ingenious but unreliable. The manufacturing precision of the seventeenth century could not consistently produce gears fine enough to mesh without jamming.

But Leibniz's most consequential contribution to computation was not a machine at all. It was an idea.

In 1679, he wrote a paper describing a complete arithmetic system that used only two digits: zero and one. He called it the binary system.

Our everyday number system is base-10 — it uses ten symbols, 0 through 9, and when you run out of symbols, you carry to the next column. We use ten because we have ten fingers. If we had eight fingers, we would use base-8. The base is a convention, not a law.

Leibniz proposed base-2. Only two symbols: 0 and 1. The number after 1 is 10 (which means "two," not "ten"). Then 11 (three), 100 (four), 101 (five). It is cumbersome for humans — the number 255 requires eight binary digits instead of three decimal ones. But Leibniz was not thinking about human convenience. He was thinking about elegance. In binary, every question reduces to the simplest possible choice: yes or no, true or false, something or nothing. Leibniz, who was deeply interested in Chinese philosophy, saw a connection between binary arithmetic and the *I Ching*, with its patterns of broken and unbroken lines. He believed he had found a number system that reflected the fundamental structure of logic itself.

He was right. But the world would not care for another 250 years. Binary arithmetic was published, admired as a philosophical curiosity, and filed away. There was no machine that could use it. Mechanical calculators worked with gears, and gears naturally represent base-10 — a wheel with ten positions is straightforward to build. Binary was a solution waiting for a problem that would not arrive until the twentieth century, when a completely different kind of machine — one built from electrical switches instead of gears — would need a number system based on the simplest possible physical states: on and off.

I will return to Leibniz in Chapter 3. For now, remember his two digits. They are patient. They will wait.

The Stick That Built the World

Before Charles Babbage, I want to spend a moment on a machine that almost no history of computing mentions, despite the fact that it was the most widely used computational tool in the world for three and a half centuries.

The slide rule was invented by William Oughtred around 1622. It is an analogue calculator. It does not count discrete units the way an abacus or a Pascaline does. Instead, it represents numbers as physical distances along a logarithmic scale, and performs multiplication and division by sliding one scale against another. Picture two rulers laid side by side, where the markings are compressed at the higher numbers. Slide one ruler along the other to the right distance, and the answer appears at the point where the scales align. No gears. No moving parts beyond the sliding motion itself. A twelve-inch stick that could multiply six-digit numbers in seconds.

For 350 years, the slide rule was the instrument of engineering and science. Every bridge designed before roughly 1975 was calculated on a slide rule. Every ship's course charted by the Royal Navy. Every artillery trajectory in two world wars. Every structural beam, every electrical circuit, every chemical process designed in the industrial age. The engineers who built the Panama Canal used slide rules. The engineers who designed the Apollo spacecraft used slide rules. An entire civilization's physical infrastructure was computed on an instrument that could fit in a shirt pocket.

The slide rule died in the mid-1970s, killed almost overnight by the electronic pocket calculator. Hewlett-Packard introduced the HP-35 in 1972 — the first scientific calculator that could fit in a pocket — and within five years the slide rule was extinct. Not declining. Extinct. Manufacturing ceased. Training stopped. A tool that had been the universal instrument of technical professionals for longer than the United States had existed disappeared in half a decade.

The slide rule's death is the purest example in computing history of what Chapter 1 called a substrate transition. The slide rule did not get worse. It did not fail at its job. It was replaced by a machine that operated on fundamentally different physics — electronics instead of mechanical sliding — and the new physics was so much faster, so much cheaper to reproduce, and so much easier to use that the old substrate became instantly obsolete. This is what a phase transition looks like: not a gradual decline, but a cliff.

The pattern will repeat with vacuum tubes, again with mainframes, and again — soon — with silicon itself.

The Engine That Could Not Be Built

Charles Babbage was, by any measure, one of the most remarkable minds of the nineteenth century. He was also, by nearly every account, one of the most difficult human beings in London — a city that was not short of difficult human beings. He quarrelled with organ grinders. He quarrelled with Parliament. He quarrelled with the craftsmen building his machines, the funders supporting his work, and the colleagues who admired him. He was brilliant, obsessive, tireless, and possessed of a temperament so abrasive that he managed to alienate nearly everyone who tried to help him.

His obsession began with errors. In the early nineteenth century, the British economy, its military, and its global navigation all depended on printed mathematical tables — logarithmic tables, trigonometric tables, astronomical tables, actuarial tables. These tables were computed by hand, by workers who were literally called "computers" — human beings whose job was to perform calculations according to fixed procedures and record the results. The work was divided into simple steps, each performed by a different person, in what amounted to an assembly line for arithmetic. The results were then typeset, printed, and distributed.

At every stage — calculation, transcription, typesetting — errors crept in. And these errors had consequences. A wrong number in a navigation table could send a ship onto rocks. A wrong number in an actuarial table could bankrupt an insurance company. Babbage, who had spent hours verifying tables and finding mistakes in nearly every volume he checked, reportedly exclaimed: "I wish to God these calculations had been executed by steam."

That wish became the Difference Engine — a mechanical device designed to compute polynomial functions automatically and stamp the results directly into copper printing plates, eliminating human error at every stage from calculation to publication. The British government, recognising the strategic value of accurate tables for the Admiralty, funded the project with £17,000 — roughly £2 million today. It was the largest government investment in a computing project in history.

The Difference Engine was never completed. Babbage's designs required gears machined to tolerances that the best craftsmen in England could not consistently achieve. Parts that should have meshed smoothly jammed. Components that should have aligned precisely drifted. Babbage, characteristically, blamed the craftsmen. The craftsmen blamed Babbage's impossible specifications. The government, having spent £17,000 with nothing to show for it, stopped funding and never got its money back.

But by the time the Difference Engine collapsed, Babbage had already conceived something far more ambitious. Something that would not be fully understood for another century.

The General-Purpose Machine

The Difference Engine was a specialist. It could compute polynomial functions and nothing else. The machine Babbage began designing in 1834 was a generalist. He called it the Analytical Engine, and it was, on paper, the first general-purpose computer ever conceived.

The parallels to a modern computer are so precise they border on the eerie.

The Analytical Engine had a *mill* — the part that performed arithmetic operations. In a modern computer, this is the processor. It had a *store* — a set of columns of gears that held numbers for use in calculations. In a modern computer, this is memory. It received instructions from punch cards — directly inspired by the Jacquard loom, a French weaving machine that used sequences of punched cards to control which threads were raised and lowered, allowing complex patterns to be woven automatically. In a modern computer, the punch cards are the program. And it had an output mechanism that could print results or punch them into cards for later use.

Processor. Memory. Program. Output. These are the four components of every computer built in the twentieth and twenty-first centuries. Babbage designed all four in the 1830s, using nothing but mechanical gears and the principle of the Jacquard loom. He even designed the Engine to support conditional branching — the ability to change its sequence of operations based on intermediate results. In modern terms, he designed "if-then" logic. In the 1830s.

The Analytical Engine was never built either. The same manufacturing limitations that defeated the Difference Engine made the far more complex Analytical Engine even more impractical. Babbage spent the rest of his life refining the design, filling notebooks with increasingly detailed plans for a machine that the technology of his era simply could not construct. He died in 1871, having built only fragments. The machine that contained every essential element of a modern computer existed for forty years as nothing but drawings and frustration.

The Poet's Daughter

In 1833, at a London dinner party, a seventeen-year-old named Ada Byron met Charles Babbage. She was the daughter of Lord Byron, the most famous poet of his era, though she had never known him — her parents separated weeks after her birth and Byron died when she was eight. Her mother, determined that Ada would not inherit her father's temperament, had her educated rigorously in mathematics and science. It was an unusual program for a young woman in the 1830s. It produced an unusual mind.

Babbage showed her the portion of the Difference Engine he had built — a working section of gleaming brass gears — and she was captivated. Not by the spectacle, which impressed most visitors as a kind of mechanical parlour trick, but by the principle. She understood, immediately, what the gears were doing. And she wanted to know what else they might do.

The relationship that followed — sustained over years of letters, conversations, and shared intellectual ambition — would produce one of the most important documents in the history of computing. In 1842, an Italian military engineer named Luigi Menabrea published a paper in French describing the Analytical Engine, based on lectures Babbage had delivered in Turin. Ada, now married and known as Ada Lovelace, was asked to translate the paper into English. She did — and then added a set of notes that were more than twice the length of the original paper.

It is these notes that secured her place in history.

In Note G, Lovelace described a detailed step-by-step procedure for using the Analytical Engine to compute Bernoulli numbers — a sequence of values important in number theory. The procedure specified the exact sequence of operations the machine should perform, the order in which cards should be fed, and the way intermediate results should be stored and reused. It is recognized today as the first computer program ever written. Lovelace wrote software for a machine that did not exist, and would not exist for another hundred years.

But her most important insight was not the algorithm. It was a conceptual leap that Babbage himself never made.

Babbage saw the Analytical Engine as a mathematical tool — a machine for computing numbers with perfect accuracy. Lovelace saw something larger. She wrote that the Engine could act upon "other things besides number," provided that those things could be expressed in relationships the machine's operations could capture. Music, she suggested, could be composed by a machine if the fundamental relationships of harmony and composition were expressed in a form the Engine could process. Patterns, logic, symbols of any kind — not just quantities — were within its reach.

This was the first articulation of general-purpose computing. Not a machine that calculates. A machine that manipulates symbols according to rules. The symbols can represent numbers, but they can also represent words, musical notes, images, logical propositions, or anything else that can be formally described. The power of the machine lies not in what it calculates but in what it can represent. Lovelace saw this in 1843, looking at a machine made of brass that had never been fully built, and described a future that would not arrive for over a century.

She also saw the limits. In the same notes, she wrote what may be the most prescient sentence in the history of artificial intelligence: "The Engine has no pretensions whatever to *originate* anything. It can do whatever we know how to order it to perform." The machine does not think. It executes. The intelligence remains with the person who writes the instructions. We will return to this sentence later, when the machines begin to do something that looks, to some observers, remarkably like originating.

> *"The Engine has no pretensions whatever to originate anything. It can do*
> *whatever we know how to order it to perform." — Ada Lovelace, 1843*

The Power in the Tables

Return to the reason Babbage started building in the first place: the tables.

The printed mathematical tables of the nineteenth century were not academic documents. They were instruments of national power. The British Admiralty's *Nautical Almanac*, published annually, contained the astronomical and navigational data that every Royal Navy captain needed to determine his ship's position at sea. Accurate tables meant accurate navigation. Accurate navigation meant ships arrived where they intended, trade goods reached their markets, and naval fleets could be coordinated across oceans. Errors in the tables meant ships ran aground, cargoes were lost, and battles were fought in the wrong place.

Britain's global empire was, in a very real sense, a computational achievement. The nation that computed best navigated best. The nation that navigated best traded best. The nation that traded best projected power farthest. The link between computation and geopolitical dominance was not metaphorical. It was as direct as the link between a printed number in a table and the course a helmsman set at dawn.

The power thread from the first chapter extends here. The shepherd who could count his flock had an advantage over the shepherd who could not. The priest who could predict an eclipse held power over those who could not. And the nation that could compute accurate navigational tables held dominion over seas that other nations could not safely cross. In every era, the machine belongs to whoever holds it. And whoever holds it extends their reach — across a pasture, across a sky, across an ocean. The reach has grown longer with every chapter of this story. In the chapters that follow, it will grow longer still.

The Ceiling

The mechanical era of computation — from Pascal's brass box in 1642 to the last slide rule manufactured in the 1970s — spanned over three centuries. In that time, humanity built machines of astonishing ingenuity. But every one of them ran into the same wall: the speed of metal.

A gear can only turn so fast before friction defeats it. A spring can only store so much energy before it breaks. A cam can only execute so many steps before its complexity exceeds what can be manufactured. A simple calculation on the Analytical Engine, had it been built, would have taken seconds or minutes. A complex computation — the kind that modern computers perform billions of times per second — might have taken weeks.

The substrate had reached its ceiling. The next leap required something that moved faster than metal, something that could switch states in millionths of a second rather than tenths, something that did not depend on the physical movement of objects through space. It required electricity.

And when electricity arrived, it would bring with it a question that Leibniz had answered 250 years earlier, though nobody had been paying attention: what is the simplest possible language a machine can speak?

Two digits. Zero and one. On and off.

The patient binary had finally found its machine.

THE ELECTRONIC REVOLUTION

The Machine Evolves

The Switch

On a twenty-one-year-old who saw that logic and wiring were the same thing, an engineer who built a secret computer with his own money and was told to destroy it, six women whose names were erased from history, and why every computer on Earth speaks a language of only two words

"We may say most aptly that the Analytical Engine weaves algebraical patterns just as the Jacquard loom weaves flowers and leaves."
— Ada Lovelace, 1843

The Switch

On the moment computation became electric, and the answer to a question most people have never thought to ask

In the autumn of 1937, a twenty-one-year-old graduate student at the Massachusetts Institute of Technology submitted a master's thesis with a title that sounded almost comically dry: "A Symbolic Analysis of Relay and Switching Circuits." The student's name was Claude Shannon. He had spent the summer working at Bell Telephone Laboratories, where the relay — an electrical switch operated by an electromagnet — was the fundamental building block of the telephone network. He had also been studying Boolean algebra, the nineteenth-century system that reduced logic itself to mathematics using only two values: true and false. Two different worlds. Two different disciplines. Nobody had connected them.

Shannon connected them.

A relay has two states: open and closed. Boolean algebra has two values: true and false. A circuit with two relays wired in series — one after the other — completes only if both are closed. That is an AND gate. A circuit with two relays wired in parallel — side by side — completes if either is closed. That is an OR gate. A relay that is normally closed and opens when current is applied is a NOT gate. Boolean logic, the abstract mathematics of true and false, could be physically built from switches and wire.

It sounds simple when stated plainly. It was not simple. George Boole had published *The Laws of Thought* in 1854 — eighty-three years earlier — showing that logical reasoning could be written as a system of equations using only two values. For eighty-three years, nobody could think of anything practical to do with it. Relays had been everywhere for decades — in telephone exchanges, industrial controls, telegraph systems. Everyone understood how they worked. But nobody had looked at a relay and seen a truth table. Nobody had looked at a logic textbook and seen a circuit diagram. Shannon, at twenty-one, saw both at the same time.

> *"Shannon saw what no one before him had seen: Boolean logic and electrical circuits were the same thing. Abstract truth and physical switches obeyed the same rules."*

This was the bridge between Leibniz's patient binary and the physical world. Leibniz had shown, in 1679, that any number could be represented with two digits. Boole had shown, in 1854, that any logical statement could be expressed with two values. Shannon showed, in 1937, that any circuit of electrical switches could implement those logical statements physically. Three insights, spanning 258 years, clicked together like the gears Babbage could never quite get to mesh.

Why Binary Won

A question almost nobody asks, despite the fact that the answer shapes virtually every piece of technology on Earth: why does every computer speak a language of only two words?

Your phone, your laptop, the servers behind every website you visit, the systems that manage air traffic and stock exchanges and hospital ventilators — all of them, without exception, represent every piece of information as sequences of ones and zeros. Every photograph you have ever taken, every song you have ever streamed, every message you have ever sent — all of it has been translated, at the machine's most fundamental level, into an astronomically long string of two symbols.

Binary is not a law of nature. It is not mathematically superior to other number systems. The number ten can be represented just as accurately in base-10, base-2, base-8, or base-16. No base is more correct than any other. So why did binary win?

The answer is not mathematics. The answer is physics. A switch is either open or closed. A vacuum tube is either conducting or not conducting. A transistor is either on or off. Two states. Two digits. One and zero. Binary won because it matched the physics of the available substrate. If the available substrate had had ten reliable states, we would use base-10 computing. In fact, there were early experiments with ternary computers — the Soviet Setun, built in 1958, used three states — but the engineering advantages of binary in electronic circuits were so overwhelming that alternatives could not compete.

Binary is a concession to atoms, not to mathematics. This will matter enormously when we encounter a machine whose fundamental unit is not a switch with two states but a quantum object that can exist in a continuous superposition of both states simultaneously. When the physics changes, the language may change with it. Binary is not forever. It is for now.

The Machine in the Country House

In January 1944, in a Victorian country estate called Bletchley Park, about fifty miles northwest of London, a machine called Colossus began operating. It was built to solve a specific and desperately urgent problem: breaking the Lorenz cipher, a German encryption system used for high-level military communications between Hitler and his generals.

The man who built it was Tommy Flowers, a General Post Office engineer. His proposal for a high-speed electronic computing machine had been rejected by the Bletchley Park leadership as impractical. Flowers built it anyway, partly with his own money. He was not a wealthy man. He was a telephone engineer who believed that vacuum tubes were more reliable than the Bletchley leadership understood, and who was willing to stake his own salary on being right.

He was right. The first Colossus used approximately 1,500 vacuum tubes. By the war's end, ten Colossus machines were in operation at Bletchley Park, the most advanced containing 2,500 tubes. They were, by a considerable margin, the most complex electronic devices in the world. Colossus could read encrypted messages from punched paper tape at five thousand characters per second — the tape flew through the reader so fast it sometimes caught fire. The intelligence produced by Colossus contributed materially to the Allied war effort. Some historians believe it shortened the war by months.

And then, at the war's end, the British government classified everything.

The Colossus machines were dismantled. The blueprints were burned. The engineers and operators were sworn to secrecy under the Official Secrets Act. Tommy Flowers was told to destroy his records. He could not tell anyone what he had built. He could not put it on his curriculum vitae. He could not publish. He could not even discuss the work with colleagues who had been in the same building. For thirty years — until the mid-1970s — the existence of Colossus was unknown to the public, to the academic computer science community, and to most of the government itself. The American ENIAC, which became operational about a year later, received credit as the first large-scale electronic computer. The British contribution was erased from the record.

Flowers spent the rest of his career as a telephone engineer. He never received significant recognition or financial reward for building the machine that helped win the war. When he applied for a bank loan in the 1950s, he could not explain the gap in his professional record or the skills he had demonstrated. The most consequential work of his life was a secret he was forbidden to tell.

"The value of Colossus was not that it could compute. It was that the enemy did not know it could compute. Those who possess advanced computational capability do not announce it."

This pattern — classification, secrecy, the deliberate concealment of computational advantage — did not end with the war. It continued through the Cold War, through the digital age, and it continues today. It will become the central concern of this book's final chapters.

The Room That Drank a City's Power

While Colossus operated in secrecy in the English countryside, a very different machine was being built in the open, at the University of Pennsylvania's Moore School of Electrical Engineering. It was funded by the United States Army, which needed a faster way to calculate artillery firing tables — the ballistic data that told gunners how to aim their weapons under varying conditions of range, wind, temperature, and elevation.

The machine was called ENIAC — the Electronic Numerical Integrator and Computer — and when it was completed in late 1945, it was a spectacle. It contained 17,468 vacuum tubes. It filled a room fifty feet long and thirty feet wide. It weighed thirty tons. It consumed 150 kilowatts of electricity — enough to power roughly a hundred American homes. There is an often-repeated story, likely apocryphal but revealing of the machine's reputation, that when ENIAC was switched on, the lights in an entire section of Philadelphia dimmed.

But ENIAC could perform five thousand additions per second. A human computer — the kind that had been calculating firing tables by hand — could perform roughly one addition every fifteen seconds. ENIAC was seventy-five thousand times faster than the human it replaced. The gap between human computation and machine computation, which had been modest in the age of mechanical calculators, had just become an abyss.

Programming ENIAC was a physical act. There was no keyboard, no screen, no programming language. To set up a new calculation, operators had to physically reconnect cables between panels — essentially rewiring the machine for each new problem. This work was done by six women: Kay McNulty, Betty Jennings, Betty Snyder, Marlyn Meltzer, Fran Bilas, and Ruth Lichterman. They had been recruited from a pool of human computers — women who calculated firing tables by hand — and were given the job of figuring out how to make the machine do the same calculations electronically. They received no manual, because none existed. They learned the machine's logic by studying its circuit diagrams and essentially teaching themselves to program by reverse-engineering the hardware.

When ENIAC was unveiled to the public in February 1946, the six women who had   d it were not introduced. They appeared in photographs but were identified only as "refrigerator ladies" — models posing with the machine for the cameras. Their names were not included in press materials. Their contribution was not acknowledged by the Army or the university. It would be decades before historians recognized them as the first professional   rs of a general-purpose electronic computer.

This is data, not aside. At every stage of computing history, the people who do the foundational work of translating human intent into machine behavior are undervalued relative to the people who design the hardware or fund the project. This pattern is as persistent as any in this book, and it has consequences that extend far beyond historical credit. It shapes who enters the field, who stays, and whose perspective informs the next generation of machines.

The Living Room in Berlin

There is one more machine from this era that deserves attention, not because it changed the world — it did not get the chance — but because it illustrates a truth from Chapter 1 that is easy to forget: technological progress can be lost.

In 1936, a young German engineer named Konrad Zuse, working in his parents' living room in Berlin, began building a programmable computing machine. By 1941, he had completed the Z3 — a device that used 2,600 electromechanical relays to perform floating-point binary arithmetic, read programs from punched film, and execute a set of instructions that included addition, subtraction, multiplication, division, and square root. Many computer historians consider the Z3 the first functional programmable digital computer.

Zuse built it with almost no institutional support. The German military, consumed by the war effort, showed intermittent interest but provided little funding. He sourced parts from wherever he could find them. His parents' apartment became a workshop. His friends helped assemble components in their spare time.

In December 1943, an Allied bombing raid destroyed the Z3 and much of Zuse's workshop. His subsequent machine, the Z4, survived the war only because Zuse evacuated it to a village in the Bavarian Alps as Berlin collapsed around him. He reportedly transported it in a military truck he talked his way into borrowing, hiding the machine among refugees fleeing the advancing Soviet army.

Zuse's work was largely unknown outside Germany during the war and for years afterward. The country was divided, its scientific community scattered, its records fragmented. By the time Zuse's contributions were widely recognized, the narrative of computing history had already been written around ENIAC, Colossus, and the Anglo-American machines. An entire branch of the evolutionary tree had been bombed, buried, and rediscovered too late to change the story as it was told.

Zuse's story is haunting for a specific reason. It is a reminder that the history of computing is not a smooth line of progress. It is a landscape shaped by war, by politics, by who gets funded and who gets forgotten, by which machines survive and which are destroyed. The Z3 was, by several important measures, ahead of its time. It did not matter. The bombs did not care.

The Ceiling of Glass

The vacuum tube era lasted, as the dominant computing substrate, for roughly fifteen years — from the mid-1940s to the early 1960s. In that time, it transformed computation from a theoretical concept into an operational reality. Machines like ENIAC, UNIVAC, and the IBM 701 proved that electronic computation worked, that it was fast, and that it could be applied to problems ranging from ballistic trajectories to census tabulation to weather prediction.

But the vacuum tube had a problem that no amount of engineering could solve: it burned out. A vacuum tube works by heating a filament until it glows, releasing electrons that can be controlled to switch current on and off. That filament, like the filament in an incandescent light bulb, has a finite lifespan. A typical tube lasted a few thousand hours before failing. In a machine with seventeen thousand tubes, this meant that on average, a tube failed roughly every few hours. The operators of early computers spent more time locating and replacing burned-out tubes than they spent running programs.

There was also the heat. Seventeen thousand tiny furnaces in a single room generated enormous amounts of thermal energy. ENIAC required an industrial air conditioning system that consumed almost as much power as the computer itself. And there was the size. Each tube was a discrete glass component that had to be individually wired into a circuit. You could not shrink a vacuum tube below a certain size without losing the vacuum or the electrical properties that made it function. A more powerful computer meant more tubes, which meant a bigger room, more wiring, more cooling, more failures, and more operators crawling through the machine with replacement tubes and soldering irons.

The vacuum tube had reached its ceiling. Like the mechanical gears before it, the glass bulb had taken computation as far as its physics would allow. To go further — faster, smaller, more reliable, less power-hungry — required a completely different kind of switch. Not a filament in a glass envelope. Something solid. Something that did not burn. Something that could be made very, very small.

On December 23, 1947, in a laboratory in Murray Hill, New Jersey, three men found it.

What This Era Made Clear

The electronic era — from Shannon's thesis through the twilight of the vacuum tube — established several truths that would govern every subsequent chapter of computing history.

The first is that the choice of physical substrate determines everything. Binary is not a mathematical necessity. It is a physical one. The two-state switch was the simplest, most reliable component available, and the entire architecture of modern computing was built to match it. When we encounter a substrate with different physical properties — and we will — the architecture will change accordingly. The substrate is not a detail. It is the foundation.

The second is that computation and secrecy are deeply intertwined. Colossus was classified because the advantage of breaking enemy codes was too valuable to reveal. ENIAC's first major task after the war was hydrogen bomb calculations. From the very beginning of the electronic era, computation was understood by governments as a strategic asset — not just a tool for science or business, but a source of military and intelligence advantage. This understanding has never changed. It has only intensified.

The third is that the people who build and program the machines are consistently undervalued relative to the machines themselves. Tommy Flowers paid for Colossus partly out of his own pocket and was told to destroy his records. The six women who   d ENIAC were unnamed in the press coverage. Konrad Zuse built the Z3 in his parents' living room with scavenged parts and had it destroyed by a bomb. The pattern is unmistakable: the human contribution to computing is perpetually discounted. Machines are celebrated. The people who make them work are footnotes.

This pattern will not improve in the chapters that follow. It will get worse. And by the time we reach the final chapters of this book, the question of who is credited, who is compensated, and who is forgotten will have expanded from a question about historical fairness to a question about whether humans are necessary at all.

And the fourth truth, the one that closes this chapter and opens the next: every substrate reaches a ceiling, and the ceiling is always physical. Gears could not spin fast enough. Vacuum tubes could not stay cool enough. The next substrate would be a crystal of semiconductor material — something that did not heat, did not burn, did not shatter, and could be made unimaginably small. It would become the most manufactured object in the history of the human species, and the story of how it came to exist begins with a piece of germanium, a strip of gold foil, a bent paper clip, and a scientist who was furious that he had not been in the room when it happened.

The Machine Evolves

The Crystal

On a tiny piece of germanium that became the most manufactured object in human history, a wounded ego that accidentally created Silicon Valley, and the approaching moment when we run out of room

"What has been achieved, what will be achieved, is nothing short of a revolution. And revolutions do not go in reverse."
— Jack Kilby, Nobel Lecture, 2000

The Crystal

On the most manufactured object in human history, and the approaching moment when we run out of room

On the afternoon of December 23, 1947, in a laboratory at Bell Telephone Laboratories in Murray Hill, New Jersey, two physicists named John Bardeen and Walter Brattain performed an experiment that would reshape the world more thoroughly than any single invention of the twentieth century. They pressed two thin strips of gold foil against a small crystal of germanium — a semiconductor, a material that conducts electricity under some conditions and blocks it under others — and observed that a small electrical signal applied to one gold contact could control a much larger current flowing through the other. The crystal could amplify.

They called it a transistor. The device was crude — a slab of crystal, two strips of foil, and a bent paper clip holding the assembly together. It looked like something a talented child might build from a science kit. It was, in function, a switch — the same kind of switch that a vacuum tube was, performing the same on-off binary operation that Shannon had mapped to Boolean logic a decade earlier. But it was a switch made of solid material instead of heated gas in a glass envelope. No filament to burn out. No vacuum to maintain. No enormous heat to dissipate. A switch that was small, cool, reliable, fast, and — most consequentially — could be manufactured in bulk.

There is a person notably absent from this moment: William Shockley, Bardeen and Brattain's supervisor. Shockley had directed the semiconductor research group with the explicit goal of finding a solid-state replacement for the vacuum tube. He was not in the room when the breakthrough happened. He had contributed critical theoretical work, but the key experimental insight belonged to Bardeen and Brattain. And Shockley, a man of ferocious intellect and equally ferocious ego, could not accept that.

The Ego and the Valley

What happened next is one of the most consequential sequences of human pettiness in the history of technology.

Shockley, determined to reclaim intellectual ownership of the transistor, essentially locked himself in a hotel room in Chicago for several weeks in early 1948 and developed an entirely different transistor design: the junction transistor, which sandwiched a thin layer of one type of semiconductor between two layers of another. It was more elegant, more reliable, and more manufacturable than Bardeen and Brattain's point-contact design. It was also, incontestably, Shockley's. He made sure everyone knew it.

All three men shared the 1956 Nobel Prize in Physics. The collaboration, such as it had been, did not survive the award. Shockley's behavior had already driven Bardeen to leave Bell Labs for the University of Illinois, where he would win a second Nobel Prize — making him one of only four people in history to win two Nobels. Brattain was reassigned within Bell Labs and spent his later career in relative obscurity.

Shockley, meanwhile, decided to commercialize the transistor himself. In 1956, he founded Shockley Semiconductor Laboratory. He chose to locate it in Mountain View, California, near Palo Alto, partly because Stanford University had a growing engineering community, and partly — by several accounts — because his aging mother lived nearby. The most consequential industrial location decision of the twentieth century was influenced, at least in part, by a son who wanted to be close to his mum.

Shockley recruited brilliantly. He assembled a team of the best young semiconductor physicists and engineers in the country. And then he managed them catastrophically. He was paranoid, abusive, and dictatorial. He subjected employees to polygraph tests. He publicly berated their work. He refused to pursue the silicon-based research his own team believed was the path forward, insisting on a quixotic pursuit of a four-layer diode that almost nobody else thought was commercially viable.

In 1957, less than a year after the company's founding, eight of Shockley's top researchers walked out. Shockley called them "the traitorous eight." They called themselves free. With funding from Sherman Fairchild, they founded Fairchild Semiconductor — and in doing so, created the seed from which virtually the entire modern semiconductor industry would grow.

Two of the eight, Robert Noyce and Gordon Moore, would leave Fairchild in 1968 to found Intel. Another, Eugene Kleiner, would co-found Kleiner Perkins, one of the most influential venture capital firms in Silicon Valley history. By one widely cited estimate, more than sixty-five semiconductor companies can trace their lineage back to Fairchild.

"The most consequential industrial cluster in modern history exists partly because one man was brilliant and impossible to work with. Silicon Valley is the geography of a defection."

I linger on this story because it illustrates something about the evolution of computing that the purely technical narrative misses. Technologies do not evolve in a vacuum. They evolve through human institutions, and human institutions are shaped by the personalities, decisions, and dysfunctions of the people who run them. Shockley's genius was real. His toxicity was also real. And the Silicon Valley that exists today — with its culture of defection, startup formation, and venture-funded risk-taking — is, in meaningful ways, a response to the experience of working for William Shockley. The valley's founding trauma shaped its founding culture.

The Switch That Does Not Burn

Set the human drama aside for a moment and consider what the transistor actually is, because its properties explain why it did not merely improve computing but fundamentally transformed it.

A vacuum tube heats a filament to release electrons. A transistor controls electron flow through a solid crystal using electrical fields. No filament means no burnout. The most common failure mode of the vacuum tube — the thing that made ENIAC's operators spend their days crawling through the machine with replacement parts — simply does not exist in a transistor.

A vacuum tube generates substantial heat. A single transistor generates almost none. This means you can pack transistors far more densely without the machine melting itself. A vacuum tube switches in microseconds — millionths of a second. A transistor switches in nanoseconds — billionths of a second. A thousand times faster. And as transistors shrink, they get faster still, because the electrons have less distance to travel.

But the property that changed the world was not speed, not reliability, not size. It was manufacturability. A vacuum tube is a discrete handmade component. Each one must be individually manufactured, tested, and wired into a circuit by a human being. A transistor can be fabricated as part of a batch process — thousands, then millions, then billions at a time, etched onto a single slab of silicon using photographic techniques.

Building a vacuum tube computer was like building a cathedral: every stone placed by a skilled craftsman, every joint fitted by hand, every finished structure unique. Building a transistor computer would become like printing a newspaper: a pattern defined once, reproduced millions of times at negligible marginal cost. This is the shift from craft to industry, from artisanal to automated, from scarce to abundant. And it is why, within a few decades, computational power would go from being the rarest resource on Earth to one of the most ubiquitous.

Putting It All on One Chip

The transistor was a revolution. The integrated circuit was the revolution's revolution.

By the late 1950s, transistors were being used widely, but each one was still a discrete component — a tiny object that had to be individually wired to other tiny objects on a circuit board. A computer with ten thousand transistors required ten thousand physical connections, each one a potential point of failure. Engineers called this the "tyranny of numbers."

In 1958, Jack Kilby, a newly hired engineer at Texas Instruments, had an idea so simple it was almost absurd: build the entire circuit — transistors, resistors, capacitors, and the connections between them — on a single piece of semiconductor material. Kilby was alone in the lab. He was a new hire who had not yet accrued vacation time, so he worked through the company holiday while everyone else was away.

He built a prototype in September 1958 using germanium. It was rough — the components were connected by tiny gold wires bonded by hand — but it worked. A few months later, at Fairchild Semiconductor, Robert Noyce independently developed a different approach to the same idea. Noyce's version used a planar process — building the circuit in flat layers on a silicon wafer, with connections formed by depositing metal traces directly onto the surface. No hand wiring at all. Everything photographic, everything reproducible, everything scalable.

Noyce's approach won, for a reason that by now should feel familiar: it matched the manufacturing substrate better. A planar process could be automated. It could be scaled. It could put not just dozens but eventually millions and then billions of components on a single chip. The integrated circuit did for transistors what the printing press did for letters — it turned individual handcrafted elements into a mass-reproducible pattern.

Kilby and Noyce both received credit. They never litigated against each other personally, though their companies fought over patents for years. Kilby received the Nobel Prize in Physics in 2000. Noyce, who died in 1990, did not — the Nobel is not awarded posthumously. Kilby, in his Nobel lecture, acknowledged Noyce's contribution. It was a more gracious outcome than Shockley's generation had managed.

The Observation That Became a Prophecy

In 1965, Gordon Moore published a short paper in *Electronics* magazine. He noted that the number of components on an integrated circuit had been doubling approximately every year since the integrated circuit was invented. He projected the trend would continue. In 1975, he revised the estimate to a doubling every two years. This observation became known as Moore's Law.

Moore's Law is not a law of physics. It is an observation about the rate of improvement in semiconductor manufacturing — an empirical trend, not a theoretical necessity. And yet it held, with remarkable accuracy, for over half a century. The reason is partly self-fulfilling: the entire semiconductor industry organized its research, its investment, and its product roadmaps around maintaining the pace. The prophecy sustained itself because the industry believed in it.

The numbers this sustained effort produced are staggering. Intel's first commercial microprocessor, the 4004, was released in 1971 with 2,300 transistors. In 1989, the Intel 486 contained 1.2 million. In 2005, a dual-core Pentium contained 291 million. In 2012, an eight-core Intel processor contained 2.3 billion. In 2023, Apple's M2 Ultra chip contained 134 billion transistors. From 2,300 to 134 billion in fifty-two years. A factor of more than 58 million.

If automobiles had improved at the same rate as semiconductors over the same period, a car today would travel at approximately six million miles per hour, get 100,000 miles per gallon, and cost less than a penny. The comparison is absurd, which is precisely the point. No other technology in human history has improved by a factor of 58 million in half a century. Nothing else is close.

The Most Manufactured Object in Human History

The total number of transistors manufactured by the human species, across all chips, all devices, all years of production, is estimated at roughly 13 sextillion. That is 13 followed by 21 zeros. For comparison, the estimated number of grains of sand on Earth is roughly 7.5 sextillion. We have manufactured nearly twice as many transistors as there are grains of sand on the planet. More than bricks. More than nails. More than any single component of any structure humans have ever built.

And each one of those transistors is a switch. On or off. One or zero. Leibniz's two digits, replicated 13 sextillion times, performing the binary logic that Shannon proved could be implemented in circuits, executing the computations that Babbage dreamed of and Lovelace foresaw, at speeds that would have struck every person in this book's first three chapters as indistinguishable from magic.

The Geography of Power

Manufacturing advanced semiconductors is one of the most complex industrial processes ever devised. The machines that fabricate modern chips — extreme ultraviolet lithography systems, built exclusively by a Dutch company called ASML — cost approximately $380 million each, weigh 180 tons, and require multiple Boeing 747 cargo flights to deliver. They project patterns onto silicon wafers using light with a wavelength of 13.5 nanometres — light so energetic it must travel through a vacuum because air would absorb it. The precision involved is often compared to hitting a golf ball on the moon from Earth, and even that analogy undersells it.

Only three companies on Earth can manufacture the most advanced chips: TSMC in Taiwan, Samsung in South Korea, and Intel in the United States. Of these three, TSMC dominates overwhelmingly — fabricating over 90 percent of the world's most advanced semiconductors. Apple, NVIDIA, AMD, Qualcomm, and dozens of other companies design their chips but depend on TSMC to build them.

TSMC's fabrication plants are located primarily on the island of Taiwan. Taiwan sits in the Taiwan Strait, 100 miles off the coast of mainland China, in one of the most geopolitically contested bodies of water on Earth. The Chinese government considers Taiwan a breakaway province. The United States has maintained a policy of strategic ambiguity regarding Taiwan's defense for decades. Military analysts on both sides of the Pacific have gamed scenarios in which a conflict in the Strait disrupts or destroys TSMC's fabrication capacity.

If that happened, the global supply of advanced semiconductors would effectively cease. Not decline. Cease. No company could replace TSMC's output in less than several years, if then. Every smartphone, every server, every AI training cluster, every military system that depends on advanced chips would be affected. The global economy would experience a shock that would make the 2020 supply chain crisis look like a minor inconvenience.

The power thread of the transistor era: the physical substrate of modern civilization is manufactured by one company, on one island, in one of the world's most dangerous geopolitical flashpoints. The fragility is literal. The machine that the transistor built — the entire global computational infrastructure — has a single point of failure, and that point is guarded by the ambiguous commitments of great powers who may not agree on its defense.

*"The physical substrate of modern civilization is manufactured by one company,
on one island, in one of the world's most dangerous geopolitical flashpoints. This
is not a metaphor for fragility. It is fragility."*

The Wall

We are running out of room.

The most advanced transistors manufactured today are approximately 3 nanometres in their smallest dimension. A human hair is about 80,000 nanometres wide. A strand of DNA is about 2.5 nanometres across. A single silicon atom is approximately 0.2 nanometres in diameter. The transistors on a cutting-edge chip are roughly 15 silicon atoms wide.

Fifteen atoms. That is the distance between the current state of the art and the point at which transistors can no longer be made smaller. At this scale, electrons begin to exhibit quantum mechanical behavior. They tunnel through barriers that should block them. They leak from one transistor to its neighbour. The binary distinction between on and off — the foundation of every computer described in this book — becomes unreliable.

The concern is not theoretical. It is an engineering reality being confronted right now. The techniques being used to push beyond 3 nanometres — gate-all-around transistors, backside power delivery, new materials layered at atomic scales — are extraordinary feats of engineering. But they are engineering around a physical wall, not through it. Each new process node costs more, takes longer, and delivers smaller improvements.

Moore's Law is not ending because the industry lost its ambition. It is ending because the substrate is approaching its physical limit. Silicon, like gears before it and vacuum tubes before that, is reaching its ceiling. Every substrate reaches a ceiling. Every ceiling triggers a search for a new substrate. The laws of physics do not negotiate.

What Comes Next

The transistor era gave us something no previous era of computing had produced: abundance. For the first time in the five-thousand-year history of computation, the machines became cheap enough, small enough, and reliable enough to escape the institution. Computation moved from government labs and corporate data centers into offices, then homes, then pockets.

That story — the personal computer, the internet, the smartphone — is the subject of the next two chapters. It is the most visible chapter of computing history, the one most people have lived through, and the one whose lessons are most immediately relevant to the question this book is building toward: what happens when computation stops getting cheaper and starts getting more concentrated?

Return one last time to that laboratory in Murray Hill, New Jersey, on December 23, 1947. Two men, a piece of germanium, a strip of gold foil, and a bent paper clip. A device so small and unimpressive that few people in the room grasped its significance. And a supervisor who was not present, whose wounded pride would set in motion a chain of defections, startups, and innovations that would create the most powerful industrial cluster in human history.

The transistor did not arrive through careful planning. It arrived through physics, craftsmanship, ego, and accident. The valley it spawned was not designed. It grew — the way organisms grow, through adaptation, competition, and the occasional mutation that turns out to be exactly what the environment required.

Thirteen sextillion copies later, the crystal shows no sign of surrendering its dominance. But the wall is visible now. Fifteen atoms wide. And on the other side of that wall, the physics changes. The switches stop being simply on or off. They become something stranger — a machine that can be both at once.

But that machine is still several chapters away. First, the transistor's abundance must do what abundance always does: escape the laboratory, reach the public, and change everything.

The Machine Evolves

The Escape

On the moment the computer stopped belonging to governments
and corporations, a hobbyist club that changed the world,
and the great redistribution of computational power that
lasted about thirty years before it reversed

"There is no reason anyone would want a computer in their home."
— Ken Olsen, founder of Digital Equipment Corporation, 1977

The Escape

On the moment the computer stopped belonging to institutions, and the great redistribution that lasted about thirty years before it reversed

In late 1974, in a small office in Albuquerque, New Mexico, an engineer named Ed Roberts was staring at the end of his company. MITS — Micro Instrumentation and Telemetry Systems — had been selling calculator kits by mail order, but Texas Instruments had entered the market with finished calculators at prices Roberts could not match. The company was $300,000 in debt. Roberts had one idea left: a computer kit, built around Intel's new 8080 microprocessor, cheap enough that electronics hobbyists might buy it. He called it the Altair 8800, named by the twelve-year-old daughter of one of his co-workers after a planet in *Star Trek*. He expected to sell a few hundred.

The January 1975 issue of *Popular Electronics* put the Altair on its cover. Within weeks, MITS received thousands of orders. Roberts could not fill them fast enough. The machine itself was, by any objective measure, useless — a blue metal box with toggle switches and blinking lights, no keyboard, no screen, no software. You   d it by flipping switches one bit at a time. It could not do anything a mainframe could not do millions of times faster. And it changed the world, because it was yours.

To understand why a useless machine mattered, you need to understand the world it entered.

The Priesthood

For the first thirty years of the electronic era, a computer was something you visited. You did not own one. You did not carry one. You went to it, in a specially constructed room with a raised floor and controlled air, and you submitted your work to the people who operated it. Then you waited.

The dominant machine of this era was the mainframe, and the dominant company was IBM. By the mid-1960s, IBM controlled roughly seventy percent of the computer market worldwide. Its machines were expensive — hundreds of thousands to millions of dollars — physically enormous, and operated by a specialized class of professionals who controlled access to computation the way medieval clergy controlled access to scripture. If you needed the computer to do something, you wrote your instructions on coding sheets, handed them to a keypunch operator, submitted your deck of cards through a window, and came back hours or days later. You did not touch the machine.

This was not an accident of technology. It was an architecture of control. Mainframes were designed to be shared resources, centrally managed, with every action logged, every user authenticated, every program approved by an administrator.

In 1964, IBM introduced the System/360 — a family of compatible computers that could all run the same software. The 360's compatibility meant that once a customer invested in IBM software, they were locked in. The switching cost was enormous. The ecosystem was self-reinforcing. The priesthood had its cathedral.

The Altair shattered the cathedral's window. For the first time in the history of electronic computing, an individual could own a computer. Not time-share on an institutional machine. Not submit punch cards to a priesthood. Own it. Put it on a desk. Program it however you wanted. No permission required. No gatekeeper. No waiting.

The Garage

The Homebrew Computer Club is one of those historical institutions whose influence is so wildly disproportionate to its appearance that the story sounds invented. It met in a garage, then a community centre, then an auditorium at the Stanford Linear Accelerator Center. Its members were engineers, students, and hobbyists who shared a common obsession with making computers small, cheap, and personal. There were no investors. No business plans. No pitch decks. Just people who wanted to build things and show each other what they had built.

Among the regular attendees was a twenty-four-year-old engineer named Steve Wozniak, who worked at Hewlett-Packard during the day and attended Homebrew meetings at night. Wozniak was, by all accounts, a gifted engineer — one of those people who can see the elegant solution to a complex problem the way a musician hears a melody. When he saw the Altair 8800, he saw not what it was but what it could become.

Over the next several months, Wozniak designed and hand-built a computer that was fundamentally different from the Altair. Where the Altair required toggle switches, Wozniak's machine had a keyboard. Where the Altair displayed blinking lights, Wozniak's machine could display text on a television screen. He called it the Apple I.

His friend Steve Jobs saw a business. Jobs had neither Wozniak's engineering talent nor his technical knowledge. What Jobs had was an instinct — nearly unerring, though often infuriating to those around him — for how technology should feel to the person using it. Wozniak built machines for engineers. Jobs wanted to build machines for everyone.

The Apple II, released in 1977, proved Jobs's instinct right. It came in a moulded plastic case — not a bare circuit board, not a metal box with toggle switches. It had a keyboard, a color display, and expansion slots. It looked like something that belonged in a home, not a laboratory.

The Killer App

In 1979, a Harvard Business School student named Dan Bricklin was sitting in a lecture, watching his professor erase and recalculate numbers on a blackboard. It was the same kind of tedious, error-prone arithmetic that had driven Pascal to build a calculator three centuries earlier. Bricklin's insight was the same as Pascal's: a machine should do this instead.

Bricklin owned an Apple II. This was not incidental. The Apple II was one of the only personal computers in 1978 with enough memory — 32 kilobytes, expandable — to hold a meaningful grid of numbers. It had a screen capable of displaying rows and columns of text. It had Wozniak's floppy disk interface, so you could save your work. And it had those expansion slots that Wozniak had insisted on — the ones that let you add more memory as your spreadsheets grew. Wozniak had built a machine more capable than it needed to be for hobbyist tinkering. That excess capability is what made VisiCalc possible. Jobs made the Apple II look approachable. Wozniak made it powerful enough to matter.

With his friend Bob Frankston, who wrote the actual code in 6502 assembly language, Bricklin built VisiCalc — the first electronic spreadsheet. It displayed a grid of rows and columns. Type numbers and formulae into the cells. Change one number, and every cell that depended on it recalculated instantly. VisiCalc was released for the Apple II in October 1979, and within a year it had become the reason people bought Apple IIs. Not the hardware. Not the color graphics. The spreadsheet. Accountants, financial analysts, and small business owners — people who had never considered owning a computer — bought one because VisiCalc ran on it.

The pattern would repeat so reliably that the industry gave it a name: the killer app. VisiCalc for the Apple II. Lotus 1-2-3 for the IBM PC. Desktop publishing for the Macintosh. Email for the internet. The hardware creates a capability. The software translates that capability into a problem that ordinary people recognize. The human need, not the technical feat, drives adoption.

And here, once again, is the echo of our oldest story. Bricklin, watching his professor erase and recalculate on a blackboard, was seeing exactly what Pascal saw watching his father grind through tax arithmetic. The problem is the same problem it has always been: the human mind imagines the answer but cannot reliably execute the calculation. The machine fills the gap. The substrate changes. The gap remains.

The Man Who Should Have Been Bill Gates

In 1974, a computer scientist named Gary Kildall, working from a converted tool shed near the Naval Postgraduate School in Monterey, California, created CP/M — Control Program for Microcomputers. It was, by any reasonable definition, the first commercially successful personal computer operating system. By 1981, CP/M ran on over three thousand different computer models. Kildall's company, Digital Research, was generating millions in annual revenue. He had also invented the concept of the BIOS — the Basic Input/Output System — the thin layer of software between the operating system and the hardware that every computer still uses today.

In 1980, IBM decided to enter the personal computer market and needed an operating system. IBM's team approached Bill Gates, whose company Microsoft at that point primarily sold a BASIC programming language interpreter. Gates did not have an operating system to offer. He directed IBM to Kildall.

What happened next has been mythologized and distorted for over four decades. The popular version — that Kildall was off flying his private plane while IBM waited in his lobby — comes primarily from accounts Gates later gave to reporters. Kildall and his associates disputed this version for the rest of his life. The documented facts are these: IBM representatives arrived at Digital Research. Negotiations between IBM and Dorothy Kildall, Gary's wife and co-founder of the company, stalled — not over a missed meeting, but over the terms of the deal. Digital Research's lawyer found IBM's non-disclosure agreement unacceptable. And on the substantive question of licensing, the two sides were fundamentally misaligned: Kildall wanted per-copy royalties — the compensation model of a creator who believed his work had ongoing value. IBM wanted a flat fee with no per-copy obligations.

IBM returned to Gates. This time, Gates did not redirect them. He acquired an operating system called QDOS — the Quick and Dirty Operating System, built by Tim Paterson at Seattle Computer Products by studying CP/M's published design — for $75,000, renamed it MS-DOS, and licensed it to IBM. Critically, Gates retained the right to license MS-DOS to other manufacturers. When IBM's PC became the industry standard and dozens of companies began producing compatible machines, every one of them needed MS-DOS. Gates had not invented the operating system. He had captured the distribution channel.

IBM did eventually reach an agreement with Kildall to offer CP/M alongside MS-DOS on the IBM PC. But IBM priced CP/M at $240 and MS-DOS at $40. The market chose accordingly. CP/M faded. Microsoft became the most valuable software company in the world. And Gary Kildall, the man who invented the personal computer operating system and the BIOS, spent the rest of his life fielding questions about the day he allegedly went flying instead of meeting with IBM. He died in 1994 at the age of 52.

I tell Kildall's story not as a cautionary tale about business strategy, though it serves as one. I tell it because it reveals something about the power dynamics of computing's distribution phase that the licensing dispute makes precise. Kildall wanted royalties — the model that says the creator's contribution has ongoing value. Gates wanted the channel — the model that says whoever controls distribution controls the market. The royalty model rewards invention. The flat-fee-plus-distribution model rewards control. In the personal computer era, control won. The inventor held the jar. The distributor held the gate. And the gate, as it turned out, was worth more than the jar.

> *"Kildall wanted royalties — the model that says the creator's work has ongoing value. Gates wanted the channel — the model that says whoever controls distribution controls the market. The gate was worth more than the jar."*

Making the Invisible Visible

In the early 1970s, at Xerox's Palo Alto Research Center — known as PARC — a team of researchers led by Alan Kay developed a radically different way for humans to interact with computers. Instead of typing commands into a blank screen, they created a visual world on the screen. Documents looked like documents. Folders looked like folders. You moved a pointer using a small device that rolled on your desk, and selected things by clicking a button. They called it a graphical user interface, or GUI.

The significance of the GUI is not aesthetic. It is cognitive. Before the GUI, interacting with a computer required you to learn the computer's language — a set of commands, each with specific syntax, that you typed into a prompt. You had to remember that "dir" listed files, that "cd" changed directories, that "del" deleted and there was no undo. The computer's internal world was invisible, and you navigated it by memorising incantations.

The GUI made the invisible visible. Files became objects you could see. Actions became gestures you could perform — dragging, dropping, clicking, resizing. The computer's internal state was represented as a spatial world that human spatial intuition could navigate. You did not need to memorize commands. You looked at the screen and saw what was there.

The GUI did not make computers faster or more powerful. In purely technical terms, a command-line interface is more efficient for an expert user. What the GUI did was lower the barrier to participation. It transformed the computer from a tool for people who spoke its language into a tool for people who did not. It was one of the largest expansions of who could compute since the invention of the abacus.

Xerox, in one of the most celebrated failures of corporate strategy in business history, did almost nothing with this invention. PARC researchers built the Alto — a complete personal computer with a GUI, a mouse, ethernet networking, and a laser printer — in 1973. Xerox management declined to commercialize it aggressively. In 1979, Steve Jobs visited PARC, saw the Alto in action, and was captivated. Apple released the Macintosh, with its GUI and mouse, in 1984. Microsoft released Windows in 1985. Xerox remained a copier company.

The Great Redistribution

What happened between roughly 1975 and 2005, because it is the most important power shift in the history of computing.

For the entirety of computing history prior to the personal computer — from the Antikythera mechanism to the IBM mainframe — computation was an institutional resource. It belonged to temples, governments, militaries, corporations, and universities. Access was controlled by gatekeepers.

The personal computer shattered this. Between 1977 and 2005, the cost of a capable computer fell from several thousand dollars to several hundred. Moore's Law put exponentially increasing power into every successive generation. The GUI made them usable by non-specialists. Software gave ordinary people compelling reasons to use them. By the early 2000s, roughly a billion people had personal computers.

A teenager in her bedroom had more computational power than the entire United States government possessed in 1960. A small business owner with a $500 computer could perform analyzes that would have required a team of accountants and a mainframe twenty years earlier. Computation had escaped the institution. It belonged, for the first time in five thousand years, to the individual.

This was the most radical redistribution of a strategic capability in human history. The power that had been concentrated in institutions since the first priests computed eclipses was suddenly, within a single generation, scattered across a billion desks.

The Recapture

The personal computer put a machine on your desk that you controlled. You chose what software to install. You stored your data on your own hard drive. You could use the machine without an internet connection, without an account, without anyone's permission. The computer was, in the most meaningful sense, yours.

The smartphone changed this. Not immediately, not explicitly, and not in a way that most people noticed. But the shift was structural, and its consequences are still unfolding.

When Apple released the iPhone in June 2007, it was marketed as a phone, a music player, and an internet communicator. It was all three, and much more. It was a computer more powerful than the desktops of a decade earlier, small enough to fit in your pocket, connected to the internet at all times, and equipped with sensors — GPS, accelerometer, camera, microphone — that made it aware of its physical environment in ways no previous consumer device had been.

It was also, in a way that the personal computer never was, tethered. The iPhone's software came from Apple. The apps you could install were approved by Apple. Your data was stored on Apple's cloud. Your identity was managed by Apple. On Android, the same was true of Google. The smartphone looked personal, felt personal, and lived in your pocket. But its fundamental architecture was a client-server relationship — your device was the client, and a platform controlled by a very large corporation was the server. The device in your hand was a window into someone else's system.

The personal computer had given you a machine whose software you chose, whose data you stored locally, whose operation required no network connection and no account. The smartphone reversed every one of these properties. Your software came from an app store controlled by the platform. Your data was stored on the platform's cloud. Your identity was managed by the platform. Your device was, architecturally, a terminal — a thin client connected to someone else's server, running someone else's software, storing your life on someone else's infrastructure. The liberation that the personal computer had achieved was being quietly unwound by a device that felt more personal than anything that had come before.

Cloud computing completed the recapture. In the personal computer era, your data lived on your machine. In the cloud era, it migrated to remote servers operated by Amazon, Google, Microsoft, and a handful of other providers. Your documents, your photographs, your emails, your financial records, your medical data — all of it moved from hardware you owned to infrastructure you rented. The economics were compelling: cloud storage was cheaper, more reliable, and accessible from anywhere. The strategic consequence was less visible: your data was no longer yours in any meaningful physical sense. It existed on someone else's servers, subject to someone else's terms of service, accessible to someone else's employees, and discoverable by someone else's government.

> *"The personal computer put computation in your hands. The smartphone put it in your pocket. The cloud put it back in someone else's building. The great redistribution lasted about thirty years."*

By 2025, three companies — Amazon Web Services, Microsoft Azure, and Google Cloud — control the majority of the world's cloud computing capacity. The pattern is clear: computation starts concentrated. A new substrate makes it abundant. The abundance distributes power. And then, through economics, network effects, and the natural tendency of platforms to consolidate, power reconcentrates in new hands.

Name this pattern: distribute, then recapture. The personal computer distributed. The cloud recaptured. The internet distributed. The platforms recaptured. Mobile distributed reach. App stores and data collection recaptured control. Every cycle of apparent democratization has been followed by a reconcentration of power in fewer, larger, more opaque institutions.

Not conspiracy. Economics. Platforms benefit from scale. Scale creates network effects. Network effects create lock-in. Lock-in concentrates power. The mechanism is so reliable that I have come to treat it as something close to a law of computational economics: abundance distributes capability, but economics recaptures control.

The Number in Your Pocket

By 2025, there were approximately 6.8 billion smartphone subscriptions worldwide, in a global population of roughly 8 billion. No previous invention — not electricity, not the automobile, not the telephone, not television — reached this fraction of the human population this quickly.

Each of those smartphones is more computationally powerful than the machines that guided the Apollo missions to the moon. Each one contains more transistors than the entire global installed base of computers in 1970. Each one can access, via the internet, more information than was contained in the Library of Alexandria.

And each one reports to a platform.

The duality of the personal computing era is plain. It distributed computational power to nearly every human being on Earth. It also created a surveillance and data infrastructure of unprecedented scope, concentrated in the hands of a few corporations and the governments that can compel them. Every search you make, every location you visit, every message you send, every photograph you take — all of it flows through infrastructure you do not control, to databases you cannot inspect, analyzed by algorithms you cannot see.

The deeper question — who really benefits from a computer in every pocket — is more complicated than the hobbyists of the Homebrew Computer Club imagined. They dreamed of liberation. They achieved distribution. Whether distribution and liberation are the same thing is a question that the subsequent chapters will increasingly force us to confront.

What Comes Next

The personal computer gave individuals access to computation. But a computer in isolation — disconnected, unable to communicate with other machines — is limited in the same way that a person who speaks but cannot listen is limited. The power of computation grows not just with the speed of a single machine but with the number of machines it can reach.

The next chapter is the story of what happened when the machines learned to talk to each other. It begins with a crashed message between two California universities, passes through a memo that a supervisor described as "vague but exciting," and arrives at the most radical transformation of human communication since the printing press — along with the most radical concentration of informational power since the invention of the secret police.

The internet was designed to be free. What it became is a different story entirely.

The Machine Evolves

The Web

On the network that connected every computer on Earth,
a first message that crashed after two letters, an architecture
designed for democracy that became an oligarchy, and the
lesson that matters most for everything that follows

*"The internet is not something you just dump something on.
It's not a big truck. It's a series of tubes."*
— Senator Ted Stevens, 2006

The Web

On the network that connected every computer on Earth, and the lesson that matters most for everything that follows

On the evening of October 29, 1969, a graduate student named Charley Kline, sitting at a computer terminal in a lab at the University of California, Los Angeles, attempted to send a message to a computer at the Stanford Research Institute, 350 miles to the north. The two machines were connected by a dedicated telephone line and a pair of newly built devices called Interface Message Processors — refrigerator-sized boxes that handled the work of chopping data into packets and reassembling them at the other end. The setup was called ARPANET, funded by the Advanced Research Projects Agency of the United States Department of Defense, and this was its first test.

Kline's task was simple: type the word "LOGIN" and see if it appeared at Stanford. He typed the L. He called the Stanford operator on a separate telephone line. "Did you get the L?" "Got the L." He typed the O. "Did you get the O?" "Got the O." He typed the G. The system crashed.

The first message ever transmitted on what would become the internet was two letters: L-O. "LO." As in: lo and behold. They fixed the bug, successfully logged in about an hour later, and the event passed without fanfare. No press coverage. No celebration. Leonard Kleinrock, the UCLA professor who supervised the lab, later noted that they had not prepared any commemorative statement — unlike, say, the team behind the first telegraph message, who had the good sense to transmit "What hath God wrought?"

This is, I have come to believe, a reliable indicator of genuinely transformative technology. The people building it are usually too close to the problem to see the magnitude of the solution. The significance arrives later, and it is always identified by people who were not in the room.

The Myth and the Reason

The most persistent myth about the internet is that it was designed to survive a nuclear war. This is half true and entirely misleading.

In the early 1960s, a researcher at the RAND Corporation named Paul Baran published a series of papers on distributed communications networks. Baran was indeed concerned about nuclear survivability — the vulnerability of centralized telephone networks to a Soviet first strike. He proposed a network architecture with no central hub: if any node was destroyed, traffic could route around the damage through surviving nodes.

But ARPANET was not built to implement Baran's proposal. ARPANET was funded by ARPA — a defense research agency, certainly, but its purpose was not nuclear survivability. Its purpose was resource sharing. In the 1960s, computers were enormously expensive, and different universities had different machines with different capabilities. ARPANET's original goal was to let researchers at one university use computers at another university remotely, without leaving their desks.

The nuclear myth persists because it makes for a better origin story. A network born to survive apocalypse sounds dramatic and inevitable. A network born to let academics share expensive machines sounds bureaucratic and contingent. But the mundane story is more revealing, because it illustrates a pattern we have seen before: the first use of a computing breakthrough is almost never its important use. ARPANET was built for resource sharing. It became the infrastructure of global communication, commerce, entertainment, surveillance, and warfare. Nobody in that UCLA lab in 1969 anticipated any of this.

The Rules of the Road

In 1983, ARPANET adopted a protocol suite called TCP/IP — Transmission Control Protocol and Internet Protocol — designed primarily by Vint Cerf and Bob Kahn. TCP/IP did something deceptively simple: it established a universal standard for how data would be packaged, addressed, routed, and reassembled across any network, regardless of the underlying hardware. A message from a computer in Tokyo could reach a computer in London, even if the two machines were completely different models running completely different software, as long as both spoke TCP/IP.

This was the internet's equivalent of the railroad gauge standardization. Before TCP/IP, computer networks were like the railroads of 1865 — dozens of incompatible systems, each optimized for local use, unable to interoperate. TCP/IP was the standard gauge. It made every network part of one network. And just as railroad standardization unleashed an explosion of commerce across the American continent, TCP/IP unleashed an explosion of communication across the world.

But there is a critical design choice embedded in TCP/IP that deserves attention. TCP/IP is decentralized. There is no master server. There is no central authority that routes all traffic. There is no single point whose failure brings down the network. Every node is, architecturally, equal. A message from a teenager's laptop and a message from the Pentagon traverse the same network using the same rules. The protocol does not know or care who is sending the message, what it contains, or how important the sender believes it to be. It treats every packet the same way.

This was a philosophical choice as much as an engineering one. The internet's architects — many of them academics steeped in the open, collaborative culture of university research — built a network that reflected their values: open, flat, egalitarian, resistant to central control. The architecture was, in a meaningful sense, democratic. No gatekeeper. No censor. No king.

Remember this. It will matter in about ten paragraphs.

The Memo

The internet and the World Wide Web are not the same thing, and the confusion between them obscures one of the most important inventions in the history of communication.

The internet is the infrastructure — the physical cables, the routers, the protocols that allow computers to talk to each other. It existed by the mid-1980s and was used primarily by academics, government researchers, and the military. It was powerful but essentially invisible to ordinary people, because using it required technical knowledge and a tolerance for cryptic addresses and error messages.

The World Wide Web is what made the internet usable by everyone.

In March 1989, a British physicist named Tim Berners-Lee, working at CERN — the European Organization for Nuclear Research, in Geneva, Switzerland — submitted a proposal to his supervisor titled "Information Management: A Proposal." Berners-Lee's problem was mundane. CERN employed thousands of researchers from dozens of countries, each using different computer systems, and the institutional knowledge was scattered across incompatible databases. He wanted a system where any document could link to any other document, regardless of where it was stored, and anyone on the network could follow those links freely.

His supervisor's response, handwritten on the cover page of the proposal, has become one of the most famous marginal notes in the history of technology: "Vague but exciting."

> *"Berners-Lee's supervisor wrote on his proposal: 'Vague but exciting.' It is possibly the most understated assessment of a world-changing idea ever committed to paper."*

Over the next two years, Berners-Lee built the three technologies that constitute the web. HTML — Hypertext Markup Language — was the format for creating documents with embedded links. HTTP — Hypertext Transfer Protocol — was the set of rules for requesting and delivering those documents. And the URL — Uniform Resource Locator — was the addressing system that gave every document on the network a unique, human-readable name.

None of these technologies was individually revolutionary. What Berners-Lee did was combine them into a coherent system, build it on top of the existing internet infrastructure, and — critically — give it away. He did not patent the web. He did not license the technologies. He released them freely, for anyone to use, extend, and build upon. This decision, more than any technical innovation, is why the web became universal. It had no owner. It had no gatekeeper. Anyone could create a website. Anyone could link to anyone else.

In 1993, a team at the University of Illinois released Mosaic — the first web browser that combined text and images in a single window and was easy enough for a non-technical person to use. The lead developer, Marc Andreessen, would later co-found Netscape, whose IPO in August 1995 — the stock nearly tripled on its first day of trading, reaching a market cap of roughly $2.9 billion — marked the moment the business world recognized that the web was not a curiosity. It was an economy.

The Explosion

What happened between 1995 and 2005 was, by any historical standard, an explosion. The number of websites grew from roughly 23,000 in 1995 to over 64 million by 2005. Global internet users grew from approximately 16 million to over a billion. Email replaced letters. Search engines replaced encyclopaedias. Streaming replaced record stores. The web did not merely add a new channel to existing industries. It restructured them.

The web spread as fast as it did because of two structural properties, both inherited from its design. The first was zero barrier to entry. Creating a website required no license, no approval, no capital investment beyond a cheap computer and a connection. A teenager could create a website that reached the same global audience as a Fortune 500 company. A blogger in Jakarta and a newspaper in New York competed on the same network under the same rules. The web was, in its first decade, the most level playing field the world of information had ever seen.

The second was the link. A link costs nothing to create and nothing to follow. Every link makes both the source and the destination more valuable. The more links, the more connections, the more valuable the network. This is what economists call a network effect — a system whose value increases with the number of participants. The telephone is the classic example: one phone is useless, two phones make a connection, a million phones make an indispensable infrastructure. The web worked the same way, except that the cost of adding a new node was essentially zero, so the network grew at a rate that no physical infrastructure could match.

But network effects have a dark twin, and it was hiding in plain sight.

The Capture

Network effects create value. They also concentrate it.

The mechanism is simple. In a network, the nodes that have the most connections attract more connections — because being connected to a well-connected node is more valuable than being connected to an isolated one. Sociologists call this preferential attachment. You might know it by a simpler name: the rich get richer. In a network with zero barriers to entry and strong network effects, the distribution of connections follows a power law — a tiny number of nodes accumulate an enormous share of the connections, and the vast majority have very few.

In the late 1990s, the web was genuinely decentralized. Millions of small websites, each independently operated. Search engines like AltaVista, Yahoo, Excite, and Lycos competed roughly equally. E-commerce was fragmented. Social interaction happened on thousands of independent forums and message boards. No single company dominated any major category.

By 2010, the landscape had consolidated into something almost unrecognisable. Google handled over 90 percent of global search. Facebook had over 500 million users and was the dominant social platform in most of the world. Amazon was the default starting point for online shopping. YouTube, acquired by Google in 2006, was the dominant video platform. A handful of platforms — fewer than ten — accounted for the majority of time spent online.

The architecture of the internet remained decentralized. The economics of the internet had become radically concentrated. The protocol still treated every packet equally. The market treated Google, Facebook, and Amazon as essential infrastructure and everything else as optional. The pipes were still democratic. The destinations were an oligarchy.

The lesson is the most important lesson in the history of the networked era: architectural democracy does not guarantee political democracy. You can build a system with no central authority, no gatekeepers, and no barriers to entry, and it will still consolidate into a small number of dominant players. The consolidation is not a corruption of the system. It is a consequence of the system's own rules — network effects, preferential attachment, and the economics of scale. The open architecture creates the conditions for the very concentration it was designed to prevent.

The internet is not the first open system in computing history to be captured by a small number of gatekeepers, and it will not be the last. If the pattern holds — and I have seen nothing in my career that suggests it will not — then every open, democratic, decentralized computing system will eventually concentrate. The question is not whether it will happen. The question is who ends up holding the power when it does.

The New Gatekeepers

The platform companies that emerged from the web's consolidation are not like the mainframe gatekeepers of the 1960s. They are more powerful, in ways that the mainframe companies never imagined.

IBM controlled access to computation. Google, Facebook, and Amazon control access to attention. In an economy where attention is the scarcest resource, controlling its allocation is a form of power that no previous technology company ever wielded. A search engine that decides what you see first when you ask a question has more influence over public opinion than any newspaper editor, any television network, any government information ministry. Not because it tells you what to think — it is more subtle than that — but because it decides what you encounter. The curation is the power.

Consider what the major platforms know. Google knows what you search for, which results you click, what you watch on YouTube, where you go via Google Maps and Android location services, what you write in Gmail, and what events are on your calendar. Facebook knows who your friends are, what you like, what you share, what you read, what you photograph, and where you check in. Amazon knows what you buy, what you browse but do not buy, what you read on Kindle, what you ask Alexa, and what you watch on Prime Video. Apple knows what apps you use, what music you listen to, what messages you send, and the biometric data your watch collects from your wrist.

No institution in human history has ever possessed this depth and breadth of information about this many individual human beings. Not the Catholic Church at the height of the confessional. Not the Soviet KGB at the height of its surveillance apparatus. Not any government, any military, any intelligence agency in the history of the world. The major technology platforms have achieved, through the voluntary adoption of consumer products, a degree of informational awareness that the most ambitious totalitarian state could only dream of — and they have achieved it without coercion, without legislation, without anyone in particular deciding that it should happen.

It happened because the services were useful. It happened because the terms of service were long and nobody read them. It happened because the cost was zero in monetary terms, and the price paid in data was invisible. It happened because network effects made the dominant platforms indispensable, and leaving an indispensable platform carries social costs that most people are unwilling to bear. It happened, in short, because the system worked exactly as designed. The design just produced consequences that the designers did not foresee and the users did not notice until it was too late to easily reverse.

"No institution in human history has possessed this depth of information about this many individual human beings. The platforms achieved it without coercion. They achieved it because the services were useful and the price paid in data was invisible."

The Vulnerability

There is one more consequence of the networked era that must be named, because it introduces a vulnerability that did not exist in any previous era of computing.

A standalone computer can only be compromised by someone who has physical access to it. You have to be in the room. You have to touch the machine. The security perimeter is the walls of the building. A networked computer can be compromised by anyone, anywhere, at any time, from any distance. The security perimeter is the entire internet. And the internet, by design, has no perimeter. It was built to connect everything to everything. Security — the act of deciding what should not connect to what — was not a design consideration. It was an afterthought.

Not a criticism of the internet's designers. They were building a research network for a community of academics who trusted each other. Security against hostile actors was not a relevant concern in 1969. But the network they built was so useful, so powerful, and so adaptable that it escaped the research community and became the infrastructure of global civilization — without ever being redesigned for that purpose. The internet that carries your banking transactions, your medical records, your military communications, and the control signals for your power grid is, at its core, the same trusting network that

Charley Kline used to send two letters to Stanford in 1969.

The security consequences have been exactly what you would expect from a system designed for trust and deployed in a world of adversaries. Data breaches exposing hundreds of millions of records. Ransomware attacks that shut down hospitals and pipelines. State-sponsored cyber operations that penetrate government agencies and critical infrastructure. An entire ecosystem of cybercrime, cyber-espionage, and cyber-warfare that exists because the foundational infrastructure of the digital economy was built without a lock on the door.

The networked era created two new forms of asymmetric power that did not exist before. The first is informational: the platforms know everything about you and you know nothing about them. The second is offensive: anyone with sufficient skill can reach into any networked system on Earth and extract, alter, or destroy its contents. The same openness that made the internet revolutionary made it vulnerable. The same architecture that resisted central control resists central defense.

The Pattern, Restated

The internet was a new substrate for computation — not in the physical sense (it ran on the same silicon that preceded it) but in the architectural sense. It transformed the unit of computation from a single machine to a network of machines. A computer connected to the internet is categorically more powerful than the same computer in isolation, in the same way that a cell in a multicellular organism is categorically more capable than the same cell floating alone.

This new substrate followed the pattern we have traced in the preceding chapters. It expanded who could compute — from institutions with mainframes to anyone with a browser. It produced an unexpected transformative application — the web, which no one anticipated when ARPANET was designed for resource sharing. It hit a governance crisis — security, privacy, platform concentration, misinformation. And it concentrated power in new gatekeepers who look nothing like the old gatekeepers but hold even more influence.

The distribute-then-recapture pattern played out in textbook fashion. The web distributed the power to publish. The platforms recaptured the power to be found. The internet distributed the power to communicate. The platforms recaptured the power to curate. Open architecture, concentrated economics. Democratic pipes, oligarchic destinations.

If this pattern holds — and I have found no reason in the historical record to believe it will not — then the question for every subsequent computing substrate is not whether concentration will occur, but how fast, how deep, and who benefits.

The next chapter is about what happens when the machines on this network begin to learn. When computation stops being something humans program and starts being something that emerges from data, patterns, and a process that even the engineers who design it cannot always explain. The story of artificial

intelligence is eighty years old. It has died three times. The fourth life looks different from the first three —
not because the ambition has changed, but because the substrate has finally caught up.

And the power implications of a machine that learns are unlike anything in the five thousand years of
computing history we have traced so far. Because a machine that calculates extends the human mind. A
machine that learns begins, however imperfectly, to resemble it.

PART IV

THE INTELLIGENT ERA

The Machine Evolves

The Neuron

On machines that learn, three deaths and four lives of
artificial intelligence, the exiles who kept believing, and
why the cost of staying at the frontier may determine
who owns the future

"I propose to consider the question, 'Can machines think?'"
— Alan Turing, 1950

The Neuron

Every machine described in this book so far has one thing in common: it does what it is told. The abacus moves beads where your fingers push them. Babbage's Engine executes the operations specified by its punch cards. ENIAC performs the calculations its operators wire into its panels. A laptop runs the software a   r wrote. The machines have grown incomprehensibly faster, smaller, and more powerful across five thousand years of evolution. But the fundamental relationship between human and machine has not changed: the human specifies. The machine executes.

This chapter is about the moment that relationship began to shift.

The idea is simple to state and extraordinarily difficult to implement: instead of telling a machine exactly what to do, show it examples of what you want and let it figure out the rules for itself. Instead of programming, teach. Instead of specifying, demonstrate. Instead of writing instructions, provide data and let the machine find the patterns.

This idea is called machine learning, and the particular approach that has come to dominate the field is inspired — loosely, imperfectly, but genuinely — by the structure of the human brain. It is called a neural network. And the story of how it went from a theoretical curiosity in 1943 to the most transformative technology of the 2020s is one of the strangest in the history of computation, because it died three times before it finally lived.

The First Life: The Perceptron

In 1943, a neurophysiologist named Warren McCulloch and a logician named Walter Pitts published a paper that proposed a mathematical model of a biological neuron. The real neuron in your brain is a cell that receives electrical signals from other neurons through connections called synapses. If the combined strength of those incoming signals exceeds a threshold, the neuron fires — it sends its own signal out to other neurons. If the incoming signals are too weak, the neuron stays silent.

McCulloch and Pitts's insight was that this behavior could be modeled as a mathematical function. Inputs come in. Each input has a weight — a number that represents how important that input is. The weighted inputs are added together. If the sum exceeds a threshold, the output is 1 (fire). If not, the output is 0 (silent). One artificial neuron. A tiny mathematical machine that takes numbers in and puts a number out.

A single artificial neuron is not intelligent. It is barely a calculator. But McCulloch and Pitts showed that networks of these simple units, connected to each other in layers, could in theory compute anything — the same *anything* that Babbage's Analytical Engine and Turing's universal machine could compute. The power was not in the individual neuron. It was in the connections.

On July 8, 1958, the *New York Times* ran a story about a machine at the Cornell Aeronautical Laboratory in Buffalo, New York. The article described it as a device that the Navy expected would eventually be able to walk, talk, see, write, reproduce itself, and be conscious of its existence. The machine was called the Perceptron. It had been built by a psychologist named Frank Rosenblatt, and it could be *trained* — not   d, trained — to recognize simple visual patterns. You showed it an image. You told it whether the image was, say, a square or not a square. And the machine adjusted the weights on its connections, strengthening the ones that led to correct answers and weakening the ones that did not. After enough examples, the Perceptron could classify images it had never seen before.

This was genuinely remarkable. For the first time in history, a machine had learned something from experience rather than being explicitly told. The weights — the knowledge — were not designed by a human. They emerged from the data. But the headline bore little resemblance to reality, and the hype it generated would prove catastrophic.

The First Death

In 1969, Marvin Minsky and Seymour Papert, two of the most respected computer scientists at MIT, published a book called *Perceptrons*. They proved mathematically that a single-layer perceptron had fundamental limitations. It could not learn certain simple logical functions. The most famous example was XOR — exclusive or — a function that outputs true when exactly one of its inputs is true, but not both. A trivially simple operation that any conventional computer could perform in a single step was provably impossible for a single-layer perceptron to learn.

Minsky and Papert's proof was correct. Their conclusion — that neural networks were a dead end — was not. Multi-layer networks could solve XOR and far more complex problems. But the book did not emphasize this possibility, and the field read the book as a death sentence. Funding agencies, looking for reasons to cut budgets, seized on the result. Government grants for neural network research dried up almost entirely. Academic positions disappeared. Researchers who continued working on neural networks found themselves marginalized within their own departments.

Frank Rosenblatt, the man who had built the Perceptron, died in a boating accident on Chesapeake Bay on July 11, 1971. He was forty-three years old. He did not live to see his work vindicated.

This was the first AI winter. Not a gradual cooling but a sudden freeze. The research did not slow down. It stopped. The community did not shrink. It scattered.

"The first AI winter was not a gradual cooling. It was a sudden freeze. The community did not shrink. It scattered. And the few who continued believed in something the field had abandoned."

The Second Life and the Second Death

In 1986, a paper by David Rumelhart, Geoffrey Hinton, and Ronald Williams demonstrated a practical method for training multi-layer neural networks. The technique was called backpropagation — a way of calculating how much each weight in the network contributed to an error, and adjusting those weights accordingly, working backward from the output to the input. The idea was not entirely new — versions of it had been described by several researchers in the 1960s and 1970s — but the 1986 paper presented it clearly enough, and demonstrated it convincingly enough, that the field paid attention.

Imagine teaching a child to throw a ball at a target. The child throws. The ball lands three feet to the left. You say "a little more to the right." The child adjusts. The ball lands one foot to the left. "A bit more." Eventually, the throws converge on the target. Backpropagation is this process, automated. The network makes a prediction. The prediction is compared to the correct answer. The difference is propagated back through the network, and every weight is nudged in the direction that would have reduced the error. Repeat this millions of times, with millions of examples, and the weights converge on a configuration that produces accurate predictions.

Backpropagation revived the field. Neural networks could now be built with multiple layers, trained on real data, and applied to practical problems like handwriting recognition and speech processing. The second life had begun.

But parallel to this revival, a different approach to artificial intelligence was absorbing most of the field's energy and nearly all of its commercial investment. They were called expert systems — programs that encoded human expertise as a set of if-then rules. If the patient has a fever AND a cough AND a rash, THEN consider diagnosis X. If the stock price drops below Y AND trading volume exceeds Z, THEN execute sell order. The knowledge was handcrafted by humans, not learned from data. Every rule was explicitly written by an expert. Expert systems were the darling of the 1980s AI industry. Companies invested hundreds of millions of dollars. Japan launched a national Fifth Generation Computer project. Corporations bought specialized hardware to run expert systems. The hype was enormous.

And once again, the technology could not deliver what the hype promised. Expert systems were brittle — they worked well within their narrow domain but failed catastrophically when they encountered situations their rules did not cover. They were expensive to build, because every rule had to be extracted from a human expert and hand-coded. They were difficult to maintain, because as knowledge changed, every affected rule had to be manually updated. By the early 1990s, the expert systems market had collapsed. Japan's Fifth Generation project was quietly wound down. The specialized hardware became expensive paperweights.

This was the second AI winter. The entire field of artificial intelligence had once again promised more than it could deliver. And for neural network researchers, the winter was doubly cold: not only had their field's broader umbrella lost credibility, but their specific approach was still regarded by the mainstream as a sideshow. The real work, the establishment believed, was in symbolic AI — hand-coded rules, logic, knowledge representation. Neural networks were messy, opaque, and theoretically dubious.

The Exile

What happened during the second AI winter is, I think, one of the most instructive episodes in the history of computing, and it connects directly to the power dynamics this book has been tracing.

A small number of researchers continued working on neural networks through the 1990s and early 2000s, largely without major funding, without mainstream academic recognition, and often in institutional settings far from the elite departments that had declared the approach a dead end. Geoffrey Hinton worked at the University of Toronto in Canada. Yann LeCun worked at Bell Labs and later at New York University. Yoshua Bengio worked at the University of Montreal. These three, later called the "godfathers of deep learning," shared a conviction that deep networks could learn representations of extraordinary complexity, if only the right training techniques and sufficient data could be brought to bear. They were right. But being right during a winter is a lonely experience.

Hinton has described submitting papers to major conferences during this period and having them rejected repeatedly. The community was not interested. The approach was unfashionable. For a decade and a half, the deep learning researchers were academic dissidents — believers in a paradigm the mainstream had dismissed, sustained by a handful of government grants, a few supportive colleagues, and the stubborn conviction that the mathematics was sound.

The pattern that will recur in the final chapters: *innovation disperses during winters and concentrates during springs*. When funding collapses, researchers scatter — to smaller universities, to foreign countries, to the margins of their field. They work independently, quietly, cheaply. The ideas diversify because there is no dominant institution enforcing orthodoxy. But when the breakthrough comes — when spring arrives — the largest institutions move immediately to capture the talent, the technology, and the strategic advantage. The springs are harvests. The winters are planting seasons. And the people who endure the winter are rarely the ones who profit most from the spring.

The Third Life

The spring arrived in 2012, and it arrived with a bang.

Every year, the computer vision community held a competition called the ImageNet Large Scale Visual Recognition Challenge. Research teams submitted systems that attempted to classify one million photographs into a thousand categories. Winning teams typically improved on the previous year's best result by a fraction of a percentage point, the way track records fall by hundredths of a second.

In 2012, a team from the University of Toronto submitted a deep neural network called AlexNet, designed by Alex Krizhevsky under the supervision of Geoffrey Hinton. AlexNet did not improve on the previous year's result by a fraction. It demolished it. The error rate dropped by over ten percentage points

— a margin of victory so large that it was as if an athlete had broken the world record by thirty seconds. The result was not incremental. It was categorical.

Two things made AlexNet possible that had not been available during the winter. The first was data. ImageNet contained over fourteen million labeled images — photographs that humans had annotated by hand. Neural networks are hungry. They learn from examples, and the more examples they see, the better they learn. Fourteen million was an ocean no previous generation had possessed.

The second was hardware. AlexNet was trained not on a conventional processor but on graphics processing units — GPUs. A GPU is a chip originally designed to render video game graphics. It excels at performing the same mathematical operation on millions of data points simultaneously — precisely what training a neural network requires. A single GPU could perform in hours the training that would have taken weeks on a conventional processor. The substrate had caught up to the algorithm.

Hinton's ideas in the 1990s were sound. His algorithms were correct. What he lacked was the data and the hardware to make them work at scale. The concept was right. The atoms — or rather, the transistors — were not yet sufficient. When the transistors caught up, the ideas that had been dormant for twenty years erupted.

The spring that followed was the most intense in AI history. Google hired Hinton in 2013. Facebook hired LeCun the same year. Bengio's lab in Montreal became a global magnet for talent. Within five years, deep learning had been applied to speech recognition, language translation, drug discovery, protein structure prediction, autonomous navigation, game playing at superhuman levels, and dozens of other domains. Each success attracted more funding, which attracted more talent, which produced more success. In 2018, Hinton, LeCun, and Bengio shared the Turing Award — computing's highest honor — for their work on deep learning.

The Transformer

In 2017, a team of eight researchers at Google published a paper with a title that was, in retrospect, breathtakingly accurate: "Attention Is All You Need."

To understand why it mattered, you need to understand the problem it solved. Language is sequential. A sentence unfolds word by word, and the meaning of each word depends on the words that came before it and sometimes on the words that come after. The word "bank" means one thing in "river bank" and a different thing in "bank account." To understand language, a machine must understand context — the relationships between words, even words that are far apart in a sentence or a paragraph.

Previous neural network architectures for language processed text one word at a time, in order, passing information forward through a chain. This was slow and lossy — by the time the network reached the end of a long sentence, it had partially forgotten the beginning. The transformer replaced this sequential processing with something called *attention* — a mechanism that allows the network to look at every word in a sentence simultaneously and learn which words are most relevant to which other words, regardless of how far apart they are.

Think of reading a mystery novel. A sequential reader processes each page in order and tries to hold everything in memory. An attentive reader, reaching a crucial clue on page 200, can flip back to page 30 and recognize its significance. The transformer is the attentive reader. It does not process language word by word. It processes it all at once, attending to the relationships between every pair of words in the input.

This architecture turned out to be extraordinarily powerful, extraordinarily scalable, and — the detail that changed the world — extraordinarily responsive to being made larger. When you train a transformer with more data, on more hardware, with more parameters, it does not just improve incrementally. It develops capabilities that were not present at smaller scales. It begins to summarize text, answer questions, write code, translate between languages, and perform tasks it was never explicitly trained to do. These emergent capabilities are among the most debated phenomena in modern computer science. Nobody fully understands why they appear.

The Fourth Life

On November 30, 2022, a San Francisco company called OpenAI released a product called ChatGPT. It was a large language model — a transformer trained on a vast corpus of text from the internet, books, and other sources — wrapped in a simple chat interface that anyone could use. You typed a question. It typed an answer. In plain English. About anything.

ChatGPT reached one hundred million users in approximately two months. No consumer technology in history had been adopted that fast. Not the telephone. Not television. Not the internet. Not the iPhone. It was not adopted because it was perfect — it was not. It made errors, fabricated facts, and could be manipulated into producing harmful content. It was adopted because, for the first time, a machine could participate in a conversation in natural language, at a level of fluency and apparent comprehension that felt, to the human on the other side, genuinely responsive.

I am choosing my words carefully here. I said "apparent comprehension" and "felt responsive." Whether a large language model actually understands anything is one of the most contested questions in contemporary science and philosophy. Some researchers argue that LLMs are performing a very sophisticated form of pattern matching — "statistical parrots," in one critic's phrase. Others argue that the distinction between "real" understanding and a sufficiently detailed model of understanding is unclear, and may not be meaningful.

I do not propose to resolve this question. But I want to reframe it, because I think the debate over machine consciousness distracts from the question that actually matters. The question that matters is not whether the machine thinks. The question is what the machine enables. The printing press did not think. It transformed civilization. The steam engine did not think. It industrialized the world. The relevant question about any tool is not what is happening inside it. It is what becomes possible because it exists. And what becomes possible because large language models exist is a transformation of human productivity, human creativity, and human access to information that is still being measured, still being debated, and still unfolding.

The Cost of the Frontier

The power thread tightens to its narrowest point so far.

Training a frontier large language model — the kind of model that produces the most capable AI systems available — requires three things in quantities that effectively exclude all but a handful of organizations on Earth.

The first is data. Frontier models are trained on trillions of tokens of text — a significant fraction of everything that has ever been written on the internet, supplemented by books, academic papers, code repositories, and other sources. Acquiring, cleaning, and preparing this data is a massive engineering operation.

The second is hardware. Training a frontier model requires thousands of the most advanced GPUs available — chips that cost tens of thousands of dollars each, manufactured almost exclusively by Nvidia using fabrication capacity at TSMC. The training process runs continuously for weeks or months, consuming enough electricity to power a small city.

The third is money. Credible estimates of the cost of training a single frontier model range from $100 million to well over $1 billion, and the costs are rising with each generation. This is not the cost of running the model once it is trained. This is the cost of creating it. And each new generation typically requires several times more investment than the previous one.

Count the organizations that can spend a billion dollars on a single experiment. The list is short. It includes a few of the largest technology companies — Google, Microsoft, Meta, Amazon, Apple — and a handful of well-funded startups backed by those same companies and their investors. It includes the national AI programs of the United States, China, and a few other nations. It does not include most universities, most companies, most governments, or any individual.

A new phenomenon in the history of computation, and it inverts the pattern we traced in earlier chapters. The personal computer era distributed computational capability because the cost fell. The AI era is concentrating computational capability because the cost of staying at the frontier is rising exponentially. The machine in your pocket is more powerful than ever. But the machine that matters most — the frontier model, the one that defines the boundary of what AI can do — is less accessible to you than any computing technology since the mainframe.

"The personal computer era distributed capability because the cost fell. The AI era is concentrating capability because the cost of the frontier is rising

What the Winters Teach

The three AI winters reveal a pattern they reveal is directly relevant to what is happening now and what will happen next.

In biological evolution, mass extinction events — the end-Permian, the end-Cretaceous — are catastrophic in the short term and creative in the long term. The Permian extinction killed roughly 90 percent of all species on Earth. But it cleared the ecological space for the dinosaurs. The Cretaceous extinction killed the dinosaurs, but it cleared the space for mammals — and eventually for us.

AI winters worked the same way. The first winter killed the Perceptron hype but cleared space for backpropagation. The second winter killed expert systems but cleared space for deep learning. The third, quieter winter of the 1990s forced the field's most stubborn believers into exile, where they developed the ideas that would explode when the substrate caught up. Each winter killed the weak ideas and the opportunistic funding. Each spring validated the ideas that had survived. And each spring was more powerful than the last, because the surviving ideas had been refined during the winters by people who continued working when there was no reward for doing so.

After each previous winter, the spring was relatively open. New companies formed. New labs contributed. The barrier to entry was low enough that a graduate student with a good idea and access to a university computer could make a significant contribution. AlexNet, the breakthrough that launched the current era, was built by a PhD student and his professor.

The current spring is different. The cost of entry has risen so dramatically that the barrier is not intellectual but financial. A brilliant researcher with a transformative idea but no access to a billion-dollar compute cluster cannot test that idea at the frontier. The spring is open in principle and closed in practice. The ideas are free. The substrate is not.

The pattern from earlier chapters — distribute, then recapture — playing out at the level of the AI substrate itself. The algorithms are published. The papers are open. The code is often available. But the ability to run the code at the scale that matters is concentrated in a handful of the wealthiest organizations on Earth. Open knowledge, closed capability. It is the internet's story retold: democratic architecture, oligarchic economics.

Lovelace's Question, Revisited

Earlier in this book, I quoted Ada Lovelace's observation that the Analytical Engine "has no pretensions whatever to originate anything. It can do whatever we know how to order it to perform."

A large language model complicates this statement in ways that Lovelace could not have anticipated. Nobody "ordered" ChatGPT to write a sonnet in the style of Shakespeare about quantum physics. Nobody d a specific rule for combining those concepts. The model learned patterns from the vast corpus of text it was trained on, and it combines those patterns in ways that its creators did not specify and often cannot predict. Whether this constitutes "origination" is a question for philosophers. That it produces outputs no human explicitly designed is a matter of observation.

I raise this not to settle the question but to mark the moment. For the first time in the five-thousand-year history of computation traced in this book, the relationship between human intent and machine output has become genuinely ambiguous. Every previous machine did exactly what it was told. This machine does something in the vicinity of what it is told, in ways that emerge from its training rather than its instructions. The human is still in the loop — the prompt shapes the output. But the machine's contribution is no longer merely execution. It is, at minimum, interpolation. Whether interpolation is a form of creativity is a debate for another book. That it changes the fundamental dynamic between humans and machines is the point that matters here.

And with that changed dynamic comes a changed power structure. When the machine merely executes, the human who gives the instructions holds all the power. When the machine contributes something of its own — however we define that contribution — the question of who controls the machine becomes more complex. And when that machine is available only to those who can afford the substrate, the answer to who controls it becomes, increasingly, the answer to who holds the future.

What Comes Next

The neural network, in its current form as the large language model, is the most capable computing technology humanity has ever produced. It is also the most concentrated. It is also the most opaque — even its creators cannot fully explain why it produces the outputs it does. And it is also the most hungry, consuming data, hardware, and energy at a scale that increases with every generation.

But the neural network, for all its power, is still running on the same substrate we have been using since 1947: transistors on silicon, switching between one and zero. The current era of AI is pushing that substrate harder than any previous application. The largest training runs consume the output of entire power plants. The demand for GPUs has created global shortages and reshaped international trade policy. The substrate is groaning.

The next chapter is about what the computer opened its eyes to see — the smartphone and its fifteen sensors, the military signal the state gave away for free, and the surveillance infrastructure that six billion people adopted voluntarily. Then, the final chapters of this book trace two substrate transitions that are arriving simultaneously: a machine that computes using the quantum behavior of subatomic particles, and a machine built from the molecules of life itself. Both raise a question that no previous substrate has forced us to confront: what happens when the next generation of computational power is so expensive, so complex, and so strategically consequential that it does not distribute at all?

The Machine Evolves

The Sensor

On the moment the computer opened its eyes, the fifteen
instruments hiding in your pocket, a military signal the
state gave away for free, and the surveillance infrastructure
that six billion people adopted voluntarily

*"If you have something that you don't want anyone to know,
maybe you shouldn't be doing it."*
— Eric Schmidt, CEO of Google, 2009

The Sensor

On the moment the computer opened its eyes, and the surveillance infrastructure that six billion people adopted voluntarily

In 2009, a German politician named Malte Spitz asked his mobile phone carrier, Deutsche Telekom, for all the data they had collected about him. They refused. He sued. After a court settlement, Deutsche Telekom delivered a spreadsheet. It contained 35,831 rows.

Each row represented a single instance when Spitz's phone had communicated with the network over a six-month period — from August 31, 2009 to February 28, 2010. Working with the German newspaper *Die Zeit* and a team of data visualization specialists, Spitz turned the spreadsheet into an interactive map. Press play, and you watch six months of a human life unfold in real time. Every train ride. Every night at home. Every trip to another city. Every time he lingered in one location or rushed through another. The map showed when he slept, when he traveled, when he worked, and where he spent Christmas. It showed who he called and how often, though Spitz chose to redact those names to protect the privacy of the people in his life.

Thirty-five thousand data points. Six months. A portrait of a human life more detailed than anything the person living it could have reconstructed from memory. And this was 2009 — before smartphones carried most of the sensors they carry today. Before GPS tracking was continuous. Before fitness trackers measured heartbeats and sleep cycles. Before facial recognition and LiDAR and always-on microphones. What Deutsche Telekom collected in 2009, with the comparatively primitive technology of that era, was the equivalent of a rough pencil sketch. What your phone collects today is a high-resolution photograph, updated every second, of every dimension of your physical existence.

Take out your phone. Hold it in your hand. It weighs roughly six ounces. You probably think of it as a device for communication and entertainment. That is how it was marketed. It is not what it is.

What you are holding is the most densely instrumented device in the history of the human species.

Fifteen Instruments

Inside that thin slab of glass and aluminium are, at minimum, the following sensors. An accelerometer, which measures how fast and in which direction the phone is moving. A gyroscope, which measures the phone's orientation and rotation in three-dimensional space. A magnetometer, which detects magnetic fields and functions as a digital compass. A barometer, which measures atmospheric pressure and can determine your altitude — which floor of a building you are on. A proximity sensor. An ambient light sensor.

A GPS receiver, which triangulates your position on Earth to within a few metres by listening to signals from satellites orbiting twenty thousand kilometres overhead. At least two cameras — front and rear —

capable of capturing still images and video at resolutions that exceed what professional film cameras could produce twenty years ago. A microphone sensitive enough to pick up a whisper across a quiet room. A fingerprint scanner or facial recognition system that maps the geometry of your body. An NFC chip for contactless payments. Bluetooth, Wi-Fi, and cellular radios that continuously scan for nearby devices and networks.

Some newer models add a LiDAR scanner that builds three-dimensional maps of the space around you using pulses of infrared light. Some include ultra-wideband chips that can locate other devices to within centimetres. Some, paired with a watch on your wrist, measure your heart rate, blood oxygen, skin temperature, and sleep patterns.

Fifteen instruments. Many of them running continuously. All of them generating data. And nearly seven billion of these devices are in active use worldwide.

The Blind Calculator

For the first sixty years of the electronic era, computers were blind, deaf, and numb. They processed information, but they did not perceive anything. An ENIAC operator typed numbers in and got numbers out. A mainframe processed punch cards. A personal computer received keystrokes and mouse clicks. Every piece of information that entered the machine was put there deliberately by a human being. The computer knew nothing about the world that the human did not explicitly tell it.

A computer in 1995 could calculate the trajectory of a satellite but did not know whether it was sitting in a building or a field. It could process a million financial transactions per second but did not know whether its operator was standing or sitting, happy or distressed, alone or in a crowd. The machine was powerful and completely disconnected from physical reality.

The smartphone changed this so completely that it is difficult, from the vantage point of 2026, to appreciate how radical the change was. The computer did not merely become mobile. It became aware. It knows where it is. It knows how fast it is moving. It knows what it can see and hear. It knows who is holding it. It exists, for the first time in the history of computation, in the physical world — not as an abstract processor but as an entity with senses.

> *"For sixty years, computers were blind, deaf, and numb. The smartphone did not merely make the computer mobile. It gave the computer senses. It made the machine aware."*

The Senses, Mapped

The parallel to biological evolution is so direct that I want to draw it explicitly, because it illuminates what is happening and what it means.

For billions of years, the earliest life on Earth had no senses at all. Single-celled organisms drifted in the ocean, reacting only to direct chemical contact. They were, in the terms of this book, blind calculators — processing information at the point of contact, with no awareness of what lay beyond their immediate surface.

Then, over hundreds of millions of years, sensory organs evolved. Photoreceptor cells appeared — primitive light-sensitive patches that could distinguish brightness from darkness. These evolved into eyes of increasing sophistication, from the simple pit eyes of flatworms to the compound eyes of insects to the camera-like eyes of vertebrates. Ears evolved from vibration-sensitive cells in fish to the complex sound-processing organs of mammals. A vast proliferation of ways to perceive the world at a distance.

The effect was not incremental. It was transformative. An organism that can see a predator approaching from fifty metres occupies a fundamentally different ecological niche than one that detects the predator only at the moment of contact. Senses do not just provide information. They provide time — time to react, time to plan, time to choose. The organism that senses becomes an organism that anticipates. And an organism that anticipates survives better than one that merely reacts.

The smartphone gave the computer this same transition. GPS is proprioception — the sense of knowing where your body is in space. The camera is vision. The microphone is hearing. The accelerometer and gyroscope are the vestibular system — balance, orientation, a sense of movement. The barometer is environmental awareness. Bluetooth and Wi-Fi scanning are something like echolocation — detecting nearby objects by sending out signals and listening for responses.

A computer with senses is not the same category of machine as a computer without them. A blind calculator is a tool you use when you choose. A sensing computer is an environment you inhabit. You use a tool. You live inside an environment.

The Signal That Made It Possible

Every location-aware service on every smartphone on Earth — every map, every ride-hailing algorithm, every geofenced advertisement, every fitness tracker that logs your morning run, every dating app that shows you who is nearby — runs on a signal that the United States military gave away for free.

The Global Positioning System was built by the Department of Defense, beginning in 1973, as a military navigation tool. A constellation of satellites, each broadcasting a precise time signal, allowing a receiver to calculate its position on Earth to within a few metres. In 1983, after the Soviet shootdown of Korean Air Lines Flight 007 — which had strayed into Soviet airspace due to a navigation error — President Reagan directed that GPS be made available to civilian aviation. In 2000, President Clinton ordered the removal of Selective Availability — the intentional degradation of the civilian signal — making full-precision GPS accessible to anyone with a receiver.

This was, by any measure, a remarkable act of public infrastructure. The state built the constellation, launched the satellites, maintained the ground stations, and distributed the signal globally, without charge. GPS created enormous public value: safer aviation, precision agriculture, emergency response, accessible navigation for billions of people. It is one of the most consequential public goods of the twentieth century.

And it is the foundational infrastructure of modern surveillance capitalism. Every one of Malte Spitz's 35,831 data points was anchored by GPS. Every location trail that allows a mobile carrier, a social media platform, or an advertising network to map a human life to a geographical grid runs on the signal the state distributed for free. The platforms did not build the constellation. They captured the value it created. The state distributed the signal. The platforms recaptured the intelligence.

The distribute-then-recapture pattern in its purest form. The personal computer distributed because it was cheap; the cloud recaptured the data. The internet distributed because the protocol was free; the platforms recaptured the attention. GPS distributed because the state gave it away; the platforms recaptured the location. In each case, the infrastructure that enabled the distribution was open, democratic, and available to all. In each case, the value was captured by the few who understood that the real product was not the signal, but the pattern of life the signal revealed.

The Bargain

The smartphone's sensor suite was not a secret. Every phone's specifications list its sensors openly. Nobody hid the GPS receiver or the accelerometer. The information was available to anyone who cared to look. Almost nobody looked.

The reason is a bargain — implicit, never formally negotiated, and deeply asymmetric. The sensors enable services that people genuinely value. GPS enables maps. The camera enables photography. The microphone enables voice assistants. Every sensor has a consumer-facing function that provides real, tangible, daily value.

In exchange, the sensors generate a continuous stream of data about the person carrying the device — where they are, where they have been, how fast they are moving, what they are looking at, what they are saying, who they are near, how they sleep, how their heart beats. This data flows to the platforms that operate the device's software: Apple, Google, the developers of every app that has been granted sensor access.

You get navigation, and the platform gets your location history. You get a voice assistant, and the platform gets a record of your questions. You get fitness tracking, and the platform gets your biometric data. You get convenience. The platform gets surveillance. Both sides receive something of value. But the values are not symmetric. You receive a service that makes your day slightly easier. The platform receives a data point that, aggregated across billions of users over years, produces a model of human behavior more detailed and more comprehensive than any ever constructed.

The Asymmetry

The preceding chapters traced how computational capability translates into power. The priest who predicts the eclipse. The navy that navigates by superior tables. The government that classifies its codebreaking machines. The company that trains the frontier AI model. In each case, the power derives from an asymmetry: one party can compute something that the other party cannot.

The sensor era introduces a new and particularly potent form of this asymmetry. It is not an asymmetry of computation. It is an asymmetry of awareness.

The platform knows where you are. You do not know where the platform's servers are. The platform knows what you search for, what you buy, who you call, how long you sleep, and where you were at 2:47 AM last Thursday. You do not know what the platform does with this information, who has access to it, how long it is retained, or what inferences are drawn from it. The platform's algorithms analyze your behavior continuously. You cannot see the algorithms. You cannot audit them. You cannot verify their conclusions. You cannot appeal their judgments.

Not the asymmetry of a tool and its user. It is the asymmetry of an observer and its subject. And the subject — you, me, the nearly seven billion people carrying sensor-laden devices — volunteered for the observation. We were not compelled. We were not coerced. We were offered maps and music and messaging and we said yes. The most comprehensive surveillance infrastructure in human history was built not by a totalitarian government but by consumer electronics companies, adopted not through mandate but through desire.

I find this more unsettling than the alternatives. A surveillance state imposed by force can, at least in principle, be resisted. A surveillance infrastructure adopted voluntarily, because it is genuinely useful and genuinely pleasant to use, offers no such clarity. You cannot resist something you enjoy. You cannot fight something you chose. The chains are services, and the prison is convenience, and most of the inmates do not recognize the architecture because the view from inside is so comfortable.

The Data That Sees You

Your phone's location data, collected over months or years, reveals where you live, where you work, where your children go to school, which doctor you visit, which church or mosque or synagogue you attend, which political rallies you go to, who you spend the night with, and how often. This is not inference. This is geometry — a device that records its GPS coordinates every few minutes produces a map of your life that is more detailed than anything you could reconstruct from memory.

Your search history reveals your fears, your desires, your medical concerns, your financial anxieties, your relationship troubles, and your secret interests — the things you would never say aloud to another person but will type into a search bar at midnight because you believe the query is private. It is not private. It is stored. It is analyzed. It is used to build a profile of your interests and vulnerabilities that is, in many cases, more accurate than your own self-assessment.

Your biometric data — your face, your fingerprint, your heart rate, your gait — is uniquely identifying and, unlike a password, cannot be changed. If your password is stolen, you create a new one. If your facial geometry is compromised, you cannot create a new face. Biometric data is permanent, personal, and irrevocable. And billions of people have handed it to technology companies in exchange for the convenience of unlocking their phone by looking at it.

Your microphone data — even when you are not actively using a voice assistant — has been the subject of persistent concern. Major technology companies have acknowledged that human contractors reviewed recordings from voice assistants to "improve accuracy," meaning that fragments of conversations conducted in the presumed privacy of your home were heard by strangers in offices you have never seen.

None of this data is collected in secret. All of it is disclosed, technically, in terms of service and privacy policies that average between five thousand and ten thousand words per document. One study estimated that if a person actually read every privacy policy they encountered in a year, it would consume roughly 76 working days. The disclosure exists. The informed consent does not.

The Instrument of Others

The sensor data collected by consumer devices does not remain exclusively with the company that collected it. It flows.

It flows to advertisers, who use it to target messages with a precision that no previous advertising medium could approach. A billboard is seen by everyone who passes it. A targeted advertisement is seen only by people whose data profile matches specific criteria — age, income, location, browsing history, recent purchases, recent searches, proximity to a competitor's store. The advertisement does not just reach you. It reaches you at the moment your data suggests you are most receptive.

It flows to data brokers — companies whose entire business model is the collection, aggregation, and resale of personal data. These companies operate largely without public awareness. Most people have never heard of Acxiom, Oracle Data Cloud, or Experian Marketing Services, yet these companies hold detailed profiles on hundreds of millions of individuals, assembled from sensor data, purchase records, public records, and other sources.

It flows to governments. Law enforcement agencies in many countries routinely purchase location data from commercial brokers, obtaining surveillance capability that would normally require a court order if collected directly. Intelligence agencies have acknowledged using commercially available data for national security purposes. The distinction between government surveillance and commercial data collection has, in practice, dissolved. The government does not need to build a surveillance apparatus. It can buy access to one that already exists.

And it flows, in ways that are difficult to trace and nearly impossible to control, into the training data of artificial intelligence systems. The large language models described in an earlier chapter were trained on vast corpora of text from the internet — text that includes personal information, private communications that were inadvertently made public, and data scraped from platforms that users believed was shared only with friends. Your words, your photographs, your public behavior — filtered through sensors, processed through platforms, and fed into machines that learn to predict and generate human behavior — are part of the substrate on which the next generation of AI is built.

The Organism in Your Pocket

When organisms first evolved eyes, the competitive landscape of life changed permanently. Before vision, predators and prey interacted only at close range. After vision, the world expanded. An arms race began that drove a massive acceleration in the complexity of life on Earth. Many evolutionary biologists believe that the development of the eye was a primary trigger of the Cambrian explosion — the period roughly 538 million years ago when virtually all modern animal body plans appeared in a geologically brief window. Vision did not just add a capability. It restructured the entire ecosystem.

The smartphone's sensor suite is doing the same thing to the computational ecosystem. Before sensors, computation operated in an abstract space. After sensors, computation operates in physical reality — aware of location, movement, sound, light, proximity, and the biometric identity of the person holding the device. The consequences are restructuring every domain it touches: commerce, communication, transportation, healthcare, security, politics, warfare, and the basic texture of daily human experience.

The phone in your pocket is not a tool. It is, in a functionally meaningful sense, an organism — a device that senses its environment, processes information, communicates with other devices, adapts its behavior based on what it perceives, and operates continuously whether or not you are actively using it. It is not alive. But it is not passive in the way that any previous computing device was passive. It is attentive. And it is paying attention, always, to you.

The Bridge to What Comes Next

The evolution of computing has been traced from a shepherd's jar of pebbles to a device in your pocket that knows where you are, what you are looking at, how your heart is beating, and what you whispered to your phone at midnight. Eight chapters. Five thousand years. Four fundamental operations — store, retrieve, compare, transform — that have never changed. And a single thread that connects them all: the machine belongs to whoever holds it, and whoever holds it sees further than those who do not.

Everything in the first eight chapters of this book has happened. The pebbles, the gears, the vacuum tubes, the transistors, the networks, the neural networks, the sensors — all of this is history. The next four chapters are about what is beginning to happen and has not yet reached its consequence.

We are approaching two substrate transitions simultaneously, and that has never happened before. The first is quantum computing — a machine that abandons the binary switch and computes using the quantum-mechanical behavior of subatomic particles. The second is biological computing — a machine built not from silicon but from the molecules of life.

Each of these transitions, on its own, would represent the most significant change in computational substrate since the transistor. Together, they suggest something that none of the previous eight chapters has prepared us for: a future in which the machine does not merely extend the human mind but begins to merge with the natural world — and in which the power dynamics we have traced through pebbles and gears and vacuum tubes and silicon reach a concentration that has no precedent in human experience.

"Over eight chapters, the jar has grown from a handful of pebbles to a global infrastructure. The next chapters ask a harder question: what happens when only a handful of institutions on Earth can build it, afford it, or even see it?"

The Machine Evolves

The Qubit

On the end of ones and zeros, a physicist who told nature
to stop pretending to be classical, a spinning coin that
has not yet landed, and the invisible arms race that
may already be underway

*"Nature isn't classical, dammit, and if you want to make
a simulation of nature, you'd better make it quantum
mechanical."*
— Richard Feynman, 1981

The Qubit

On the end of ones and zeros, and the invisible arms race that may already be underway

In 1981, Richard Feynman stood at a blackboard at MIT and said something that sounded, to the physicists in the room, like a complaint, and to everyone who came after, like a prophecy.

The problem was simulation. Feynman wanted to simulate quantum-mechanical systems — the behavior of atoms, molecules, subatomic particles — on a computer. But a quantum system with fifty particles requires tracking an exponentially large number of possible states — more states than any classical computer could store, let alone process. The mathematics was clear. The physics was mocking him. Every time he tried to model nature on a classical machine, the machine choked on the complexity that nature handled effortlessly.

His conclusion was characteristically blunt: nature isn't classical, dammit. If you want to simulate nature faithfully, you need a computer that operates by the same rules nature does. You need a quantum computer.

Everything you have read in this book so far is about machines that speak two words: one and zero, on and off, yes and no. The abacus. The Analytical Engine. Colossus. ENIAC. Every transistor in every chip in every device on Earth. The shepherd's pebble was binary in its own way — present or absent, in the jar or not. For five thousand years, every computing substrate has been built on a foundation of discrete states. Something is, or it is not.

This chapter is about a machine that refuses this premise.

The Spinning Coin

A quantum computer does not process ones and zeros. It processes something for which there is no everyday analogy that is fully accurate, but for which the closest useful approximation is this: imagine a coin spinning in the air. While it spins, it is not heads and it is not tails. It is, in a real and physical sense, both possibilities at once. Only when it lands — only when you look — does it become one or the other.

A classical bit is a coin that has landed. Heads or tails. One or zero. A qubit — a quantum bit — is a coin in the air. It exists in a *superposition* of states, carrying information about both possibilities simultaneously, right up until the moment of measurement. It is not that we don't know which state the qubit is in. It is that the qubit is, according to the laws of quantum mechanics, genuinely in both states at the same time. The physics is unambiguous on this point, even though human intuition rebels against it.

Earlier in this book, I told you that binary was a concession to atoms, not to mathematics, and I asked you to remember that sentence. Here is where it pays off. Binary won because 1940s switches had two reliable states — on and off. But qubits do not have two states. They have a continuous range of states, represented mathematically as points on a sphere called the Bloch sphere. A qubit is not a simpler switch. It is a richer one — a switch that carries exponentially more information than a classical bit. And a machine built from these switches does not compute faster in the same way. It computes differently.

What Quantum Computing Actually Is

Quantum mechanics is the most accurately tested physical theory in the history of science, and it describes a reality that contradicts virtually every intuition you have about how the world works. I cannot make it intuitive. Nobody can. Feynman himself is widely quoted as saying: "If you think you understand quantum mechanics, you don't understand quantum mechanics." What I can do is describe the three properties that make quantum computing powerful, as clearly and honestly as the subject allows.

The first is *superposition*. A qubit can exist in a combination of 0 and 1 simultaneously. Two qubits can represent four states simultaneously. Three qubits, eight. Ten qubits, 1,024. Fifty qubits can represent over one quadrillion states simultaneously. The number of states grows exponentially with the number of qubits. This is not a faster classical computer. It is a different kind of mathematical object entirely.

The second is *entanglement*. Two qubits can be placed in a state where they are correlated — measuring one instantly determines the state of the other, regardless of the physical distance between them. Einstein famously called this "spooky action at a distance" and refused to believe it was real. It is real. It has been experimentally confirmed thousands of times. Entanglement allows quantum algorithms to explore correlations across an enormous space of possibilities in a coordinated way, rather than checking each possibility one at a time.

The third is *interference*. When multiple quantum states are superposed, they can reinforce some outcomes and cancel others, the way two waves in a pond can combine to create a bigger wave or cancel each other to create still water. A quantum algorithm is, in essence, a carefully designed sequence of operations that arranges the interference so that wrong answers cancel out and the right answer is amplified.

This is a different kind of computation entirely. A classical computer checks possibilities sequentially, or at best in modest parallel. A quantum computer uses superposition to represent all possibilities simultaneously, entanglement to coordinate information across them, and interference to filter for the correct answer. The result, for certain classes of problems, is a machine that can solve in minutes what a classical computer could not solve in the lifetime of the universe.

To be precise: *for certain classes of problems*. A quantum computer is not faster at everything. It is not a better laptop. It will not load your email faster or render your video games more smoothly. It is powerful for specific kinds of problems — problems with a particular mathematical structure that quantum algorithms can exploit. And those specific problems happen to include some of the most consequential computational challenges on Earth.

"A quantum computer does not search through possibilities one by one. It uses the physics of interference to make the correct answer more probable than any incorrect one. For certain problems, it solves in minutes what classical machines could not solve in the lifetime of the universe."

The Path Here

In 1985, David Deutsch at Oxford formalized Feynman's intuition, describing the theoretical framework for a universal quantum computer. In 1994, Peter Shor, a mathematician at Bell Labs, discovered an algorithm that would reshape the strategic landscape of the twenty-first century.

Shor's algorithm is a method for factoring large numbers into their prime components. This sounds abstract. It is not. The security of nearly every encrypted communication on Earth — your banking transactions, your private messages, military communications, diplomatic cables, corporate secrets — relies on a mathematical assumption: that factoring very large numbers is computationally impractical. A 2048-bit encryption key is secure because factoring it on the best classical supercomputer would take longer than the current age of the universe. This is not security through secrecy. It is security through mathematics — the assumption that the maths is simply too hard to break.

Shor proved that a sufficiently powerful quantum computer could factor these numbers efficiently. Not in the age of the universe. In hours. Possibly minutes.

The publication of Shor's algorithm in 1994 changed quantum computing from a theoretical curiosity into a strategic imperative. Every intelligence agency on Earth understood what it meant: a working, large-scale quantum computer would render all current public-key encryption obsolete. Every secret protected by mathematical difficulty would become readable. Not eventually, in some distant future. As soon as the machine was built.

Where We Stand

As of 2026, quantum computers exist. They are real machines, built by IBM, Google, Microsoft, IonQ, Rigetti, Quantinuum, and others, and available, in limited form, through cloud services. In 2019, Google claimed "quantum supremacy" — demonstrating that its 53-qubit Sycamore processor could perform a specific calculation in 200 seconds that would take the world's most powerful classical supercomputer approximately ten thousand years. IBM disputed the classical estimate. The debate continues.

But the current generation are, in practical terms, prototypes. Qubits are extraordinarily fragile — disturbed by the slightest vibration, temperature fluctuation, or electromagnetic interference. They lose their quantum state — a process called decoherence — in microseconds or milliseconds. Every quantum computation is a race against decoherence, and most of the engineering effort in the field is devoted to extending the time a qubit can hold its state long enough to be useful.

The spinning coin is powerful because it carries information about both states while it spins. But the coin wants to land. The physical world conspires to make it land — air resistance, vibration, the table it's spinning on. The engineer's job is to keep the coin spinning long enough to extract the answer before it lands. Current quantum computers can keep the coin spinning for a very short time. Not long enough to break encryption. Not long enough to simulate complex molecules with commercial accuracy. But long enough to prove the principle works.

The engineering path forward involves quantum error correction — using many physical qubits to create a single, more reliable "logical" qubit, the way you might use many noisy voices singing the same note to produce a clearer tone than any individual voice. Current estimates suggest that breaking RSA-2048 encryption would require roughly four thousand logical qubits, which in turn might require millions of physical qubits, depending on the error rate. The largest quantum computers today have roughly one to two thousand physical qubits. The gap is significant but not, most experts believe, permanent.

We are, in the terms of this book, roughly where classical computing was in the ENIAC era. The principle works. The engineering is brutal. The machines fill rooms, require extreme environmental control, and break constantly. But the trajectory is clear, the investment is enormous, and the people working on the problem are among the most talented physicists and engineers alive.

What It Will Do

When quantum computers reach sufficient scale and reliability — a milestone most researchers estimate is ten to twenty years away for the most demanding applications — they will be capable of things that no classical computer, regardless of its speed, can achieve.

Molecular simulation. The behavior of molecules is governed by quantum mechanics. Simulating a molecule's behavior on a classical computer requires approximations that become exponentially more costly as the molecule grows larger. A quantum computer could simulate molecular interactions with native fidelity — because the computer and the molecule obey the same physics. This means designing drugs by modeling how a candidate molecule interacts with a protein target, computationally, with a precision that no classical simulation can match. It means designing new materials — room-temperature superconductors, better battery chemistries, more efficient solar cells — by simulating their atomic structure rather than testing thousands of variations in a laboratory.

Optimization. Many of the hardest problems in logistics, finance, and engineering are optimization problems — finding the best solution among a vast number of possibilities. Route planning for delivery networks. Portfolio allocation across thousands of securities. Scheduling for airlines, hospitals, and military operations. These problems grow exponentially harder as the number of variables increases, and classical computers solve them with approximations and heuristics. Quantum algorithms can, for certain problem structures, find optimal or near-optimal solutions exponentially faster.

Cryptography. Shor's algorithm breaks all current public-key encryption. But quantum mechanics also enables a new kind of encryption — quantum key distribution — that is provably secure based on the laws of physics rather than the difficulty of a mathematical problem. An eavesdropper who intercepts a quantum-encrypted message physically disturbs it in a way that is detectable. You cannot copy a quantum state without altering it. This means that quantum computing does not just break the old lock. It provides a fundamentally new one. But the new lock is available only to those who have the quantum infrastructure to use it.

Artificial intelligence. Quantum algorithms may accelerate specific machine learning operations — optimization of neural network weights, sampling from complex probability distributions, training on data with quantum correlations. The implications are still theoretical, but if quantum computing provides even a modest speedup in AI training, it would amplify the concentration dynamics already at work: the organizations with quantum hardware would train better models, faster, widening the gap between the frontier and everyone else.

The Invisible Arms Race

In 1944, Colossus was the most powerful computer in the world, and the British government classified its existence for thirty years. The machine's value lay not just in what it could compute but in the fact that the enemy did not know it could compute. The asymmetry of knowledge — we can read your messages; you do not know we can read your messages — was more valuable than any single piece of intelligence the machine produced.

Now project this forward.

A nation or entity that achieves a breakthrough in quantum computing — specifically, a machine capable of running Shor's algorithm against current encryption standards — would possess, for the period before the rest of the world catches up, the ability to read every encrypted communication on Earth. Every diplomatic cable. Every military order. Every financial transaction. Every corporate negotiation. Every personal message that anyone believed was private.

Would they announce this capability? The history of computing, traced through every chapter of this book, gives us the answer: they would not. Colossus was classified for thirty years. The NSA's capabilities have been consistently underestimated by the public until forced disclosures revealed their scope. Every entity that has achieved a significant computational advantage has treated that advantage as a secret to be protected, not a capability to be disclosed.

This creates a problem that is fundamentally different from the nuclear arms race, and in some respects more dangerous. Nuclear weapons are visible. You can see a uranium enrichment facility from a satellite. You can detect a nuclear test from the other side of the planet. The infrastructure required is large, distinctive, and observable. This visibility is what makes deterrence possible and arms control treaties enforceable.

Quantum computing is invisible. You cannot see a qubit from a satellite. A quantum computation produces no detectable emission — no seismic signature, no radiation, no electromagnetic pulse. The infrastructure — a cryogenic cooling system, a shielded chamber, control electronics — fits inside a single building and looks, from the outside, like any other research laboratory. There is no "quantum test" that alerts the world to a breakthrough. A nation could achieve the ability to break global encryption and the rest of the world might not know for years — until the consequences start appearing in negotiations that always seem to go one way, in financial markets that move in patterns that benefit the same actors, in military deployments that consistently anticipate the adversary's plans.

Harvest Now, Decrypt Later

There is a strategy, widely discussed in the intelligence community and known informally as "harvest now, decrypt later," that makes the timeline even more urgent than the engineering milestones suggest.

The strategy is simple. Every encrypted communication that passes through the internet can be intercepted and stored — not read, because it is encrypted, but stored. Intelligence agencies have the capacity to capture and archive vast quantities of encrypted traffic. They cannot read it today. But if a quantum computer capable of breaking that encryption becomes available in ten or fifteen years, they will be able to decrypt the archive retroactively.

Which means that every encrypted communication happening right now — every diplomatic negotiation, every corporate merger discussion, every military planning session, every private message between heads of state — has an expiration date on its secrecy. Not the date the message was sent. The date someone builds the machine that can read it.

The United States National Institute of Standards and Technology has been leading a multi-year effort to develop and standardize post-quantum cryptography — encryption algorithms believed to be resistant to quantum attack. The urgency of this effort reflects an assumption that the "harvest now, decrypt later" threat is real and that the transition to quantum-resistant encryption must happen before the quantum computers arrive, not after. The transition is underway. It is also slow, expensive, and incomplete. Every day that passes between now and its completion is a day when communications are being intercepted and stored by entities betting they will be able to read them eventually.

The Binary Question, Answered

A question has been running through this book since Leibniz proposed his two-digit system and Shannon showed how binary maps to switches. Here is the payoff.

Binary is not a permanent feature of computation. It was an engineering choice, made because switches have two reliable states, and it served us extraordinarily well for eighty years. But a qubit is not a switch. It does not have two states. It exists on a continuous sphere of possibilities, defined by two parameters — an angle and a phase — that together describe a point on the Bloch sphere. A qubit carries more information than a bit, not because it is faster, but because it occupies a richer mathematical space.

And as the next chapter will show, DNA stores information in four bases — A, T, G, C — effectively base-4 rather than base-2. The biological substrate of computation does not speak binary either.

The eighty-year reign of the binary switch may turn out to be a historical parenthesis — a solution perfectly adapted to one particular substrate, as dominant in its era as the slide rule was in its own, and potentially just as transient. The next substrates compute in the language that their physics demands, and that language is not ones and zeros. It is richer, stranger, and more powerful.

The implication is striking: it means that the most fundamental assumption of modern computing — that everything reduces to bits — may not survive the century. The foundation that Shannon built, that Turing formalized, that the entire digital economy rests on, is an artefact of a particular era's physics. When the physics changes, the foundation changes with it. And that means everything built on that foundation — every operating system, every programming language, every algorithm, every architecture — will eventually have to be rethought for a post-binary world.

The Power Question, Sharpened

Building a useful quantum computer requires infrastructure that is beyond the reach of individuals, most companies, and most nations. The machines operate at temperatures near absolute zero — colder than outer space. They require electromagnetic shielding that blocks every stray signal. They require control electronics of extraordinary precision. They require a supply chain of specialized components that no single company produces. They require physicists, engineers, and mathematicians whose training takes a decade and whose numbers could fit in a lecture hall.

The cost of a single useful quantum computing installation is measured in billions of dollars. The number of entities on Earth that can spend billions on a computation experiment is vanishingly small. This is not a garage technology. It is not a hobbyist's project. It is a sovereign capability — the computational equivalent of a nuclear program, accessible only to the wealthiest nations and the largest corporations.

And unlike the personal computer, which distributed because it was cheap, or the internet, which distributed because the protocol was free, quantum computing has no obvious path to mass distribution. The physics itself resists it. Qubits must be isolated from the environment. The more you isolate them, the more expensive and complex the apparatus becomes. There is no Moore's Law for qubits — no observed trend of exponential improvement that suggests quantum computers will become cheap, small, and ubiquitous. They may become available through cloud services, as some already are. But cloud access is not ownership. Cloud access is access granted at the discretion of the owner.

This means that the distribute-then-recapture pattern traced through the preceding chapters may not apply. There may be no distribution phase. The quantum computing era may begin concentrated and stay concentrated. The jar may never leave the hands of the few who can afford to build it.

What Comes Next

If quantum computing represents the extension of computation into the subatomic world — exploiting the physics of particles too small to see — then the next chapter represents something even stranger: the extension of computation into the living world.

Biological computing — DNA storage, molecular logic gates, engineered cells that process information — does not just use a different material. It uses a material that can grow, repair itself, copy itself, and evolve. A silicon chip must be manufactured by a billion-dollar fabrication plant. A DNA molecule can be replicated by the same cellular machinery that copies your genome every time a cell divides.

If quantum computing worries me because of its concentration, biological computing worries me for the opposite reason: its potential to propagate. A quantum computer is expensive and confined. A biological computing agent could, in principle, spread. The governance implications are unlike anything we have encountered in this book — because the governance frameworks humanity has built for computation are designed for machines. And a biological computer is not, in any traditional sense, a machine.

The spiral of computing's evolution started with nature — Jacquard mimicking the weaver's hand, Babbage modeling human calculation, McCulloch and Pitts abstracting the neuron. Then it diverged into pure engineering: electronics, silicon, integrated circuits. Now it is curving back. Quantum computing exploits the physics that governs atoms. Biological computing exploits the chemistry that governs life. The machine is returning to the substrate it originally tried to transcend.

Whether this is a homecoming or an invasion depends entirely on who is steering.

The Machine Evolves

The Molecule

On a Harvard geneticist who stored his entire book in a vial
smaller than a fingertip, one gram of DNA that could hold
every piece of data on Earth, a computer that can copy itself,
and the moment the machine stops being separate from the living world

"Life is a process which may be abstracted from other media."
— John von Neumann

The Molecule

On one gram of DNA that could hold every piece of data on Earth, and the moment the machine stops being separate from the living world

In 2012, a Harvard geneticist named George Church held up a small vial in his laboratory. The vial was smaller than a fingertip. It contained synthetic DNA. And encoded in the sequence of nucleotide bases within that DNA was his entire book — 53,000 words, plus images — stored not in silicon, not on paper, not on magnetic tape, but in the molecules of life itself.

One gram. That is how much synthetic DNA you would need to store approximately 215 petabytes of data. A petabyte is a million gigabytes. 215 petabytes is roughly the estimated total data storage capacity of every major technology company on Earth combined — every email, every photograph, every video, every financial record, every medical file, every line of code, every piece of information that the digital economy has produced and preserved.

All of it. In one gram. A quantity so small it would barely cover a fingertip.

That number sounds like science fiction. It is chemistry. DNA is the most information-dense storage medium known to exist — not because it was engineered but because it was evolved, over 3.8 billion years, by a process that relentlessly optimized for exactly the properties that data storage requires: density, durability, copyability, and fidelity.

In the preceding chapters, the computing substrate was something humans built from scratch: gears from metal, vacuum tubes from glass, transistors from purified silicon. This chapter is about a substrate that existed before humans did. A substrate that nature spent billions of years perfecting. And a substrate that raises a question that no previous computing chapter has had to confront — not just what the machine can do, but what happens when the machine and the person holding it become, at the molecular level, the same thing.

The Four-Letter Library

DNA uses a different encoding from binary. Instead of two symbols, it uses four: the nucleotide bases adenine (A), thymine (T), guanine (G), and cytosine (C). These four molecular letters are strung together in sequences that can be billions of characters long. Your genome — the complete set of instructions for building and operating your body — is approximately 3.2 billion base pairs long and is contained in nearly every cell you possess.

If binary is an alphabet with two letters, DNA is an alphabet with four. Each position in a DNA sequence carries twice as much information as a single binary digit. But the density advantage goes far beyond the alphabet. A single transistor on a modern chip is about three nanometres. A single base pair in a DNA molecule occupies about 0.34 nanometres. DNA stores information at a scale roughly ten times smaller than the best silicon technology, and it does so in three dimensions rather than two — the double helix is a coiled, twisted, folded structure that packs an extraordinary amount of information into an extraordinarily small space.

And DNA is durable. Under the right conditions — cool, dark, dry — DNA can survive for tens of thousands of years. We have extracted and sequenced readable DNA from woolly mammoths frozen in Siberian permafrost for over forty thousand years. No manufactured storage medium comes close. A hard drive lasts perhaps five to ten years. Magnetic tape, perhaps thirty years. DNA, properly preserved, lasts millennia.

"DNA is the most information-dense storage medium known to exist. One gram can hold 215 petabytes. It survives for tens of thousands of years. No manufactured medium comes close on either measure."

Writing Data to Life

The process of storing digital data in DNA works like this. Take a digital file. Convert the binary sequence into DNA bases using a mapping such as: 00 = A, 01 = T, 10 = G, 11 = C. Synthesize an actual DNA molecule with that sequence. You now have a physical molecule that encodes your digital file. To read the data back, sequence the molecule and convert back to binary.

Church's 2012 experiment was the proof of concept. In 2017, researchers at Columbia University and the New York Genome Center stored a full operating system, a movie, a computer virus, a Pioneer plaque image, and a $50 Amazon gift card in DNA and retrieved them with perfect accuracy. In 2019, researchers at the University of Washington and Microsoft demonstrated a fully automated system that could write data to DNA and read it back without human intervention. The writing process took approximately 21 hours for a small data set.

Those timescales reveal both the promise and the current limitation. Twenty-one hours is glacially slow compared to silicon. Nobody is going to stream a movie from a DNA molecule. The speed gap is a fundamental difference in physics: chemical bonds form in milliseconds; electronic switches operate in billionths of a second.

But for archival storage — data written once and read rarely, stored for decades or centuries — DNA's advantages are overwhelming. A warehouse of magnetic tape that costs millions to maintain and must be migrated every decade could be replaced by a vial that fits in a desk drawer, requires no power, and will be readable in a thousand years. The economics are converging. The cost of synthesising DNA has fallen by roughly a factor of ten thousand since the early 2000s.

The Computer That Was Always There

DNA storage is remarkable, but it is ultimately a use of biology as a passive medium — a very good hard drive. The far more radical idea is using biology as an active computer — a substrate that does not just store information but processes it.

This idea is less exotic than it sounds, because living cells have been doing exactly this for 3.8 billion years.

Consider what a single cell does. It stores information — the genome, 3.2 billion base pairs. It retrieves information — specific genes read at specific times in response to environmental signals. It compares information — the immune system distinguishes self from foreign, a feat of pattern recognition operating continuously across trillions of cells. And it transforms information — the genetic code translated into proteins that fold into three-dimensional structures carrying out the physical work of the organism.

Store. Retrieve. Compare. Transform. The four operations that opened this book — the operations that define what a computer is, from a shepherd's jar of pebbles to a quantum processor. A cell performs all four. It has been performing all four since before multicellular life existed. The cell is, by the functional definition established at the beginning of this book, a computer. It is the oldest computer on Earth. And by several important measures, it is the best.

The human brain contains approximately 86 billion neurons connected by roughly 100 trillion synapses. It performs the computational equivalent of billions of operations per second while consuming approximately 20 watts of power — roughly enough to run a dim light bulb. A modern GPU performing comparable operations consumes thousands of watts. The brain is, by a factor of hundreds or thousands, the most energy-efficient computer known.

It is also self-repairing. When a transistor fails, it stays failed. When a neuron is damaged, the brain can reroute its connections. It is self-assembling — a brain assembles itself from a single fertilized cell in approximately nine months. And it is self-replicating — the instructions for building a brain are copied every time an organism reproduces.

The question that a growing number of researchers are now asking is not whether biological systems can compute. They obviously can. The question is whether we can learn to program them.

Programming Life

In 1994, Leonard Adleman at the University of Southern California demonstrated the principle of DNA computation — encoding a graph problem into DNA sequences, mixing them in a test tube, and using biochemistry to find a solution. Billions of molecules, each representing a different potential path, reacting in parallel. It was to modern computing what the Antikythera mechanism was to the smartphone — a demonstration separated from utility by an enormous engineering gap.

But the field has advanced considerably. Researchers have built logic gates from DNA — molecular structures performing AND, OR, and NOT operations on chemical signals, just as Shannon's relay circuits performed the same operations on electrical signals. In 2024, a team from Caltech, UC Davis, and Maynooth University published a landmark result: a reprogrammable DNA computer capable of running 21 distinct algorithms using the same physical molecular system.

Living cells have been engineered to function as biosensors and to perform logical decisions: if chemical A is present AND chemical B is absent, THEN produce protein C. These engineered cells are being explored as smart drug delivery systems — therapeutic agents that circulate in the body and activate only when they detect the chemical signature of a disease.

As this book was being completed, the field crossed another threshold. In February 2026, researchers published a framework for allosteric DNA computing in *Science Advances* — a system that integrates sequence programmability with conformational dynamics, successfully bridging conformational signaling with genetic control regulations inside living biological architectures. This is not a theoretical proposal. It is complex, multi-level signal processing executed in vivo. The biological computer is no longer a laboratory curiosity. It is operating inside living systems, processing information through the same molecular machinery that the cell uses to regulate itself.

Neuromorphic chips — hardware designed to mimic the structure of biological neural networks, not just their mathematics — represent yet another convergence. Intel's Loihi chip and IBM's TrueNorth are circuits whose physical architecture mirrors biological neurons. They are dramatically more energy-efficient for certain tasks, precisely because they exploit the same principles biology discovered billions of years ago.

The Spiral

The pattern this chapter reveals is the deepest pattern in this entire book.

The history of computation is not a line from simple to complex. It is a spiral.

The spiral began in nature. The first computers were biological — living cells processing genetic information, organisms navigating their environments, brains modeling the world through neural networks of staggering complexity. Nature was computing for billions of years before humans arrived.

Humans, when they built their own machines, began by mimicking biology. Jacquard's loom mimicked the weaver's hand. Babbage modeled the human calculator. McCulloch and Pitts abstracted the biological neuron. Then the spiral diverged. Electronics, silicon, integrated circuits — purely engineered substrates with no biological counterpart. For eighty years, the machine evolved away from its origins.

And now the spiral is curving back. Quantum computing exploits the physics that governs atoms in biological molecules. DNA storage uses the actual biological medium. Neuromorphic chips mimic the architecture of neurons. The machine, after eighty years of divergence, is converging back toward the substrate it originally drew inspiration from.

This convergence is not merely technical. It is existential — and the evidence has been hiding in the oldest stories humanity tells.

The oldest written narrative in human history is the Epic of Gilgamesh, composed in Mesopotamia around 2100 BCE — in the same river valleys where a shepherd first placed pebbles in a clay jar to count his flock. In that story, a king watches his closest friend die and is so shattered by grief that he undertakes a journey to find eternal life. He crosses the waters of death. He finds a plant at the bottom of the sea that restores youth. And while he sleeps on the journey home, a serpent steals it. He returns to his city mortal. Nature reclaims the answer.

The oldest computing story and the oldest immortality story come from the same civilization. I do not think this is a coincidence. Both are stories about human beings reaching for a power that exceeds their natural capacity — to count beyond memory, to live beyond the body. For five thousand years, the two ambitions traveled separate paths. The computing ambition produced gears, engines, vacuum tubes, transistors, silicon, qubits. The immortality ambition produced mythology, alchemy, and eventually medicine.

This chapter is where the paths converge. The biological substrate that can store data, compute logic, and deliver therapies inside the human body is the same substrate that is now being used to edit the mechanisms of ageing itself. Stem cell therapies that regenerate damaged tissue. Gene editing tools — CRISPR and its successors — that correct the mutations driving age-related disease. DNA repair mechanisms borrowed from organisms that exhibit negligible senescence. Telomere extension research. Senolytics — drugs that selectively destroy the senescent cells whose accumulation drives ageing. These are not speculative. They are funded, published, and in clinical trials.

The spiral that began with a shepherd counting pebbles is arriving at a place where the machine and the body are no longer separate substrates. The same tools that program a cell to deliver a drug can program a cell to repair a chromosome. The same biological computing infrastructure that this chapter has traced from Adleman's test tube to Caltech's reprogrammable DNA computer is, in parallel laboratories, being used to extend the human lifespan. The two oldest human ambitions — to count and to live — are converging in the same molecule.

Who Gets the Nectar

In Hindu mythology, the Samudra Manthan tells the story of the churning of the cosmic ocean. The gods and the demons, locked in a struggle neither could win alone, agreed to cooperate — to churn the primordial sea together, using a mountain as a churning rod and a serpent as a rope, in order to produce Amrita: the nectar of immortality. The ocean yielded poisons and wonders as it churned. But the prize they were churning for was the nectar. The substance that would defeat death.

When the Amrita finally appeared, the demons lunged for it. They had done half the churning. They believed they had earned their share. But Vishnu, taking the form of Mohini — an enchantress so beautiful that the demons forgot their purpose — distributed the nectar exclusively to the gods. The cooperation that produced the prize was real. The distribution of the prize was a deception. Immortality was given to those who already held power and withheld from those who had helped create it.

This story was told thousands of years ago in the Indian subcontinent. It is one of the oldest narrative frameworks in human civilization. And it is the precise structural template for the governance problem that the biological substrate introduces.

Every governance framework humanity has ever built rests on an assumption so fundamental that it is never stated: power is temporary because the holder is mortal. The pharaoh dies. The emperor dies. The CEO retires or is buried. The dictator's grip weakens as the dictator's body weakens. Even the most entrenched dynasty eventually produces an heir who is less capable, less ruthless, or less interested than the founder. Mortality is nature's term limit. It is the one form of succession that no amount of power can override. It is, in the deepest sense, the mechanism by which civilization renews itself — not through choice, but through biology. The old order dies. Literally. And something new grows in the space it leaves behind.

Now imagine that the holder of the jar — the person or institution that controls quantum computation, artificial intelligence, and the biological substrate described in this chapter — uses that control to extend their own lifespan by decades. By a century. By enough to outlive every constraint that mortality would have imposed. They do not just hold the jar longer. They hold it long enough to reshape every institution, every market, every legal framework around their continued holding of it. They outlive their regulators. They outlive the journalists investigating them. They outlive the social movements resisting them. They outlive the prosecutors building cases against them. The democratic assumption that power rotates because people die ceases to apply to the people with the most power.

And consider the mathematics of the planet they inhabit. Earth's carrying capacity — its food systems, its fresh water, its arable land — evolved for populations that die. If biological interventions collapse the death rate among those who can afford them, while birth rates remain positive — even as they decline globally, they will vastly outpace deaths in a population that is no longer dying on schedule — the arithmetic becomes unsustainable. The planet becomes, in the language of population biology, a fruit fly experiment: a closed system in which reproductive output exceeds the carrying capacity because the balancing mechanism of death has been selectively disabled.

Selectively. That is the word that matters. The therapies will not be free. They will not be universally distributed. They will cost what only the holders of the jar can afford. And so the question of whether to share the nectar — the question the Samudra Manthan posed thousands of years ago — acquires a dimension that even the myth did not anticipate. Sharing the therapies universally may not be survivable at planetary scale. Withholding them is the most consequential act of gatekeeping in human history — a jar that literally contains life and death, held by people who have decided, by the act of holding it, that they deserve to live on a different timescale than the rest of the species.

But a biological computing agent introduces governance challenges beyond life extension that are categorically different from any previous computing substrate. A silicon chip does not copy itself. A DNA molecule does. A silicon chip does not grow. An engineered cell can. A silicon chip does not mutate. A biological computing agent, placed in a biological environment, is subject to the same evolutionary pressures that govern all life — mutation, selection, drift. It can change in ways its designers did not intend and cannot fully predict.

The governance frameworks that humanity has built for computing are designed for machines. Machines have serial numbers. They can be tracked, counted, embargoed, and destroyed. They stay where they are put. They do not evolve. They do not breed. A biological computing agent, once released, is subject to none of these constraints. It is governed not by export controls but by ecology. Not by patents but by evolution. Not by firewalls but by cell membranes.

The potential benefits are enormous. A drug delivery system that computes inside the body could transform medicine. DNA storage could preserve human knowledge for millennia. Engineered biological systems could perform environmental remediation. But the dual-use problem is severe: the same techniques that design a therapeutic agent can design a pathogen. The barrier between biological computing and biological weapons is not a technical barrier. It is an intentional one.

Three Dangers

The previous chapter described the danger of quantum computing: concentration. A technology so expensive that it may never distribute. The jar that never leaves the hands of the few.

This chapter describes two more.

The second danger is propagation. A technology whose defining property is that it copies itself. The jar that cannot be contained.

The third danger is permanence. A technology that allows the holder of the jar to defeat the one constraint that has, for five thousand years, guaranteed that the jar eventually passes to someone else. The holder who does not die.

The person who holds the computational jar and the biological jar simultaneously does not merely hold power. They hold power without expiration. They become, in a functional sense that would have been recognizable to every civilization that told stories about the nectar and the serpent and the churning of the ocean, the closest approximation of what humanity has historically called a god. Not because they are omniscient. Not because they are omnipotent. But because they have removed the one constraint that separated every previous holder of power from divinity: the certainty that they would die and the jar would pass to someone else.

For five thousand years, the serpent always won. Gilgamesh returned to Uruk mortal. The demons were tricked out of the Amrita. Every mythology that imagined the defeat of death also imagined its reclamation — because every storyteller lived in a world where death was, in the end, non-negotiable. The stories were warnings dressed as fantasies: you may reach for it, but you will not keep it.

The biological substrate described in this chapter is the first technology in human history that may allow the holder of the jar to keep it. And every governance framework we possess — every constitution, every term limit, every antitrust statute, every democratic norm of succession — was designed for a world in which the serpent always wins.

> *"For five thousand years, the serpent always won. The biological substrate described in this chapter is the first technology that may let the holder of the jar keep it. And every governance framework we possess was designed for a world in which the serpent always wins."*

What Comes Next

The full arc of computing's physical evolution has been traced: from pebbles to gears to vacuum tubes to transistors to silicon chips to qubits to DNA. Ten chapters. Five thousand years. Four operations. One trajectory that began in the human need to count sheep and now arrives at the ability to program molecules, extend lives, and — if the trajectory is not governed — create holders of the jar who never have to let go.

The remaining chapters of this book are not about the machines. They are about us.

The next chapter asks the question that this entire book has been building toward: what happens when a machine becomes powerful enough to model every possible outcome of a decision — and someone owns it? It is not a question about technology. It is a question about human nature. And it is, I believe, the most important question that the evolution of computation has ever posed.

Because the danger is not that the machine wakes up. The danger is that the human holding it never has to sleep.

POWER & WARNING

The Machine Evolves

Chapter 11 of 14

The Oracle Problem

On what happens when a machine can see every possible future
— and someone owns it. Three cities. Three altitudes.
One capability that arrives differently depending on where you stand.

*"Nearly all men can stand adversity, but if you want to test
a man's character, give him power."*
— attributed to Abraham Lincoln

The Oracle Problem

On what happens when a machine can see every possible future — and someone owns it

This chapter begins with you.

Not the abstract "you" of a thought experiment — the actual you, wherever you are reading this, in the specific country you woke up in this morning, governed by the specific laws that apply to you, dependent on the specific infrastructure that delivers your electricity, your water, your internet, your healthcare, your money. The Oracle does not arrive in the abstract. It arrives in your city, through your systems, under your government's jurisdiction — or outside it. Where you live determines not whether the Oracle affects you but how.

I have lived in three countries. Each gave me a different view of the same capability. What follows is not a thought experiment. It is three mornings — three ways a single technology arrives depending on where you stand when it does.

> *"The Oracle does not arrive in the abstract. It arrives in your city, through your systems, under your government's jurisdiction — or outside it. Where you live determines not whether the Oracle affects you but how."*

The Oracle in New York

You wake up in Manhattan. You work in technology. The Oracle — the convergence of quantum computing, biological computation, and artificial intelligence that the previous two chapters described — is not an abstraction to you. It is a product roadmap. It is a slide deck. It is a capability your company might build, a tool your employer might license, a competitive advantage you are paid to evaluate.

The Oracle arrives in New York as an opportunity. You can see the demos. You can attend the conferences. You can read the papers the week they are published — sometimes before, because you know the researchers. You can invest in the companies building the capability, or you can join them. You have agency. You have information. You have options. The question that occupies you is competitive: will your company build it before the other one does? Will your country maintain its lead? Will the next breakthrough happen in your lab or in Shenzhen?

The regulatory environment confirms your position. The AI companies that build the Oracle are headquartered in your city, your state, your country. They employ your neighbors. They lobby your representatives. The governance frameworks being discussed — the executive orders, the proposed legislation, the voluntary commitments — are written in your language, debated in your media, and shaped by your political process. You may disagree with the policy. But the policy is about you, written for your context, and you have a voice in it.

Your concern is legitimate and real: the pace of development may outrun the pace of governance. The companies may accumulate power faster than the institutions can constrain it. The Oracle may arrive before the rules do. But your concern operates within a system that acknowledges your existence. You are in the room. You may not control the outcome, but you are visible to the people who do.

This is the Oracle at the altitude of the builder. You can see the machine being assembled. You can touch the parts. You may even hold a wrench.

The Oracle in Toronto

You wake up in Toronto. You have universal healthcare. Your banking system is stable. Your democracy functions. Your country is prosperous, educated, and safe. By almost any measure, you live in one of the best-governed nations on Earth.

The Oracle does not arrive in Toronto as a product you build. It arrives inside the infrastructure you already depend on. Your hospital's diagnostic AI runs on an American cloud. Your bank's fraud detection uses an American model. Your military's communications depend on an American satellite constellation — one whose owner has already determined the trajectory of a major European war through unilateral connectivity decisions. The Oracle does not knock on your door. It is already in your house, woven into the systems you trust.

Your government can negotiate. Canada has PIPEDA — federal privacy legislation that imposes requirements on how data is collected, stored, and used. It has data residency provisions that require certain categories of information to remain on Canadian soil. It has a Prime Minister who stood before the World Economic Forum and named the rupture in the rules-based order — who said, publicly, that middle powers must act together because if you're not at the table, you're on the menu.

But consider what negotiation means in practice. Your hospital's diagnostic AI was trained on data from American and European populations. It uses standard diagnostic categories — categories developed for populations whose genetic backgrounds, dietary patterns, environmental exposures, and disease trajectories may differ fundamentally from the populations it now serves.

In Scarborough — the most ethnically diverse region of Toronto, and the site of Canada's largest regional dialysis program — a nephrologist named Dr. Tabo Sikaneta conducted a study of 2,397 patients with chronic kidney disease. His findings, published in *BMJ Open*, documented something that the diagnostic categories had been hiding: dialysis prevalence among immigrants was 4.2 times higher than in the Canadian-born population. Not marginally higher. Not within the range of statistical noise. Four point two times. And the disparities ran along lines that standard ethnic categories could not capture — birth country predicted outcomes more powerfully than ethnicity did. Patients born in the Caribbean, Southeast Asia, and South Asia had the highest rates. Kidney function declined fastest in Caribbean-born and South Asian patients. Diabetes prevalence, blood pressure, and glycaemic control varied dramatically by country of origin.

These patients had universal healthcare. They had equal access to the same hospitals, the same doctors, the same diagnostic tools. And they were dying at four times the rate. The reason is not that the system was withholding care. The reason is that the system was measuring the wrong things. The diagnostic categories — the standardized screens, the risk calculators, the clinical decision support tools that increasingly run on AI — were designed for a population that does not look like Scarborough. When you build an algorithm on categories that ignore birth-country epigenetics, dietary patterns shaped by migration, and the physiological consequences of systemic displacement, the algorithm will mathematically enforce the blind spot. It will not see what the categories do not measure. And it will declare the unmeasured population healthy, right up until they arrive at dialysis.

Not a failure of technology. It is a failure of measurement that technology amplifies. The algorithm did not create the blind spot. The diagnostic framework did. The algorithm inherited it, scaled it, and made it operate at the speed and confidence of a machine — which is to say, without doubt, without hesitation, and without the clinical intuition that a human physician might bring to a patient whose presentation does not fit the textbook. And this is happening in Toronto — one of the best-funded, most diverse, most data-rich healthcare systems on Earth. If the algorithm fails here, in a universal system with robust data infrastructure and PIPEDA protections, what happens when the same categories are exported to Lusaka, where the diagnostic framework was designed for a population an ocean away?

But the negotiation is asymmetric. Canada does not build frontier AI models. It does not manufacture the chips they run on. It does not operate the satellite constellation its military depends on. It can set conditions on the Oracle's use within its borders, but it cannot set conditions on the Oracle's development — because the development happens elsewhere, in labs funded by capital your country cannot match, using compute infrastructure your country does not own. The terms of the negotiation are set by the builder. The middle power accepts what is offered, negotiates at the margin, and calls the result sovereignty.

Your concern is not whether the Oracle will affect you. It already does. Your concern is whether you can govern what you cannot build. Whether privacy legislation written for the previous era of technology will hold against a capability designed in a jurisdiction that does not share your values. Whether the data about your citizens — their health records, their financial transactions, their communications — will be processed by systems whose operators answer to a different government's laws, or to no government's laws at all.

This is the Oracle at the altitude of the negotiator. You can see the machine. You cannot touch the parts. You are at the table, but you did not design the table and you cannot leave it.

The Oracle in Lusaka

You wake up in Lusaka, Zambia. You are twenty-six years old. You have a university degree — the first in your family. You work for a small agricultural cooperative that helps farmers in the Central Province access markets for their crops. Your phone is your office. It is a $90 Android device that runs on a mobile network operated by a South African telecommunications company, which contracts its cloud infrastructure from an American hyperscaler. Your money moves through a mobile banking platform whose fraud detection, credit scoring, and transaction monitoring are powered by machine learning models trained on data from three continents — none of which is yours.

The Oracle does not arrive in Lusaka as a product or a negotiation. It arrives as a condition. It is already inside the mobile money system that handles 70% of the transactions in your economy. It is inside the health screening app the NGO distributed to the rural clinics. It is inside the agricultural pricing tool that tells your farmers what their maize is worth today — a price determined by an algorithm running on a server in Virginia, trained on commodity data from Chicago, calibrated to a global market whose dynamics your farmers have no visibility into and no influence over.

Nobody asked. That is the sentence that defines the Oracle's arrival in Lusaka. Nobody asked whether you wanted predictive analytics applied to your creditworthiness. Nobody asked whether the health screening algorithm was validated on a population that looks like yours. In Toronto — with universal healthcare, world-class hospitals, and the most diverse patient population in Canada — Dr. Sikaneta found that immigrants were dying of kidney disease at four times the rate, because the diagnostic categories did not capture what was killing them. If the algorithm fails in Toronto, what does it do in Lusaka, where there is no Dr. Sikaneta, no BMJ publication, no data infrastructure to even detect the failure? Nobody asked whether the agricultural pricing model accounts for the specific conditions of Zambian smallholder farming — the soil, the rainfall patterns, the transport costs, the market access that differs from village to village. Nobody told you the data was leaving the country. Nobody told you which country it was going to. Nobody told you what decisions were being made about you, by whom, using what criteria, subject to what laws.

You cannot negotiate with the Oracle because you cannot see the Oracle. You interact with an app on your phone. The app connects to a platform. The platform connects to a cloud. The cloud connects to a model. The model was trained on data you did not contribute to, validated against benchmarks that do not include your population, and governed by policies written in a language you do not speak, in a jurisdiction that does not know you exist. Your government has no AI legislation. It has no data protection authority with the resources to audit the systems operating within its borders. It has no seat at any table where the rules for these systems are discussed. It has no mechanism of oversight, no institution with the resources or authority to audit what is being done.

Your concern is not sovereignty. You cannot lose what you never had. Your concern is visibility — whether anyone in the system even knows you are affected. Whether the model that scored your creditworthiness was designed to account for your existence. Whether the algorithm that screens your health was trained on data that includes people who look like you, live where you live, eat what you eat, carry the genetic and environmental risk factors that your population carries. The answer, in most cases, is no. You are in the margin. And the margin is where the cost is paid.

This is the Oracle at the altitude of the invisible. You cannot see the machine. You cannot touch the parts. You are not at the table. You do not know there is a table.

The Pattern at Three Altitudes

The same capability. Three experiences. The Oracle is not a hypothetical machine in a philosopher's thought experiment. It is the convergence of quantum computing, biological computation, and artificial intelligence, operating on the global data infrastructure that six billion smartphones generate, at a scale that only the largest organizations on Earth can build and operate. It is not coming. It is arriving — unevenly, at different speeds, through different channels, experienced at different altitudes depending on where you stand.

To make this concrete, because the abstract version is too easy to dismiss.

Imagine an AI system — a large language model trained on the entirety of published biomedical literature, every protein database, every clinical trial, every patent filing in pharmaceutical chemistry. This system can generate novel molecular structures the way ChatGPT generates sentences: by predicting what comes next, informed by everything that has come before. It proposes a candidate molecule for a drug that targets a specific protein involved in a specific disease.

Now connect that AI to a quantum computer. The quantum machine simulates the candidate molecule's interaction with the target protein — not approximately, the way a classical computer would, using statistical shortcuts that miss critical atomic-level behaviors, but with native quantum fidelity, because the computer and the molecule obey the same physics. The simulation takes minutes instead of months. It reveals that the candidate binds to the target with remarkable precision. The AI adjusts, proposes a variant, the quantum computer simulates again. In hours, the system has iterated through thousands of candidates that would have taken a traditional pharmaceutical lab years to evaluate.

Now connect the output to a biological computing system — a DNA synthesizer that can write the gene sequence for the winning molecule, insert it into an engineered cell, and produce the compound. The designed-by-AI, validated-by-quantum, manufactured-by-biology drug exists as a physical substance within days of the initial query.

This is the Oracle as a technical architecture, not a metaphor. Three substrates, each doing what it does best: AI generates possibilities, quantum validates them against physical reality, biology manufactures the result. The loop is closed. The time from question to molecular reality collapses from years to days.

Now change one variable. Instead of targeting a disease protein, target a protein essential to human immune function. Instead of designing a therapeutic, design a pathogen. The same architecture. The same three substrates. The same closed loop. The same collapse of timeline from years to days. The only difference is the intent of the operator. The Oracle does not know the difference. It does not care. It computes what it is asked to compute, by whoever holds it, for whatever purpose they determine.

There is a constitutional dimension to this that must be named, because it connects to the deepest legal principles the democratic world has built.

The presumption of innocence — enshrined in Article 11 of the Universal Declaration of Human Rights and the foundation of the Fifth, Sixth, and Fourteenth Amendments to the United States Constitution — holds that a person cannot be punished for something they have not done. An autonomous weapon that selects and engages a target based on a probability model is, by the standards of international humanitarian law, a war criminal by default. Not because it is malicious. Because it cannot exercise the human judgment that the Geneva Conventions require for distinction — the ability to differentiate between a combatant and a civilian — and for proportionality — the judgment that the military advantage of a strike justifies the civilian harm it causes. A probability is not a judgment. A confidence interval is not discernment. An algorithm that calculates a 70% likelihood that a building contains a combatant and authorizes a strike has treated the 30% — the civilians, the children, the people who happened to be in the wrong building — as acceptable statistical noise. This is antithetical to the legal framework that governs the use of force. Any autonomous weapon, by the fact of its autonomy, represents a breach of constitutional and humanitarian ethics.

The logic is the same logic that powers a recidivism algorithm in an American courtroom. The system that says "this person is 60% likely to reoffend, so we keep them locked up longer" is inverting the presumption of innocence — punishing a human being for something they have not done, based on characteristics they share with a demographic group. The system that says "this building is 70% likely to contain a combatant, so we strike it" is the same inversion at a different altitude. Both treat actual human beings as probabilities. Both are constitutionally indefensible. Both are operationally unchallenged — because the people affected by recidivism algorithms are disproportionately Black and poor, and the people affected by autonomous targeting are disproportionately brown and far away. The constitutional violation is hiding in plain sight because the victims lack the power to challenge it.

In New York, the Oracle is a product. In Toronto, it is a dependency. In Lusaka, it is a condition. The person in New York has agency. The person in Toronto has a negotiating position. The person in Lusaka has an app on their phone and no idea what is behind it.

The distribute-then-recapture cycle operating not across time but across geography. The capability distributes — mobile banking reaches Lusaka, diagnostic AI reaches Toronto's hospitals, the cloud reaches everywhere. But the control does not distribute with the capability. The control remains where the models are trained, where the chips are manufactured, where the data centers operate, where the capital concentrates. The further you are from the center, the less visible you are to the system — and the less the system is designed to account for you.

Every reader of this book lives at one of these three altitudes. The reader in São Paulo recognizes Lusaka. The reader in Seoul recognizes Toronto. The reader in London sees somewhere between

Toronto and New York depending on the domain. The reader in Lagos, Dhaka, and Lima knows exactly which altitude they occupy. The Oracle does not need to be explained to them. They are already living under it. They just did not know it had a name.

What the Oracle Sees

The previous two chapters documented the substrates. Chapter 9 described a quantum computer that can explore every possible solution simultaneously — a machine that does not search for the answer but computes all answers at once, collapsing to the correct one when measured. Chapter 10 described a biological computer that can store 215 petabytes in a gram of DNA, can copy itself, can evolve, and operates on an architecture that has been debugging itself for 3.8 billion years.

Combine these substrates with the data infrastructure that already exists — the six billion smartphone sensors documented in Chapter 8, the financial transaction records, the health records, the social media activity, the satellite imagery, the genomic databases, the climate data — and the Oracle is not a single machine. It is a capability that emerges when sufficient compute meets sufficient data at sufficient scale.

What does that capability look like? It looks like this: a system that can predict, with a precision that no previous technology approached, the health trajectory of a population, the creditworthiness of a borrower, the likelihood that a defendant will reoffend, the probability that a student will drop out, the risk that a business will default, the yield of a crop, the spread of a disease, the outcome of an election, the movement of a market, and the behavior of a person. Not because it knows the future. Because it has enough data about the past, enough compute to process it, and enough statistical power to identify patterns that no human analyst could detect.

Not science fiction. Fragments of the Oracle already exist. A healthcare algorithm that processed 200 million patients predicted their future utilization — and confused cost with need, systematically underserving Black patients. A sentencing algorithm predicted recidivism — and encoded racial bias as mathematics. A social media recommendation engine predicted engagement — and optimized for outrage. Each was a partial Oracle — a system that could see a fragment of the future and act on it. Each was also wrong in ways that the people who built it did not anticipate and that the people who lived under it could not detect.

The Oracle Problem is not what happens when prediction fails. It is what happens when prediction succeeds — when the machine's forecast is accurate enough to be actionable, and the entity that owns the machine acts on it. Because the action changes the future. And the entity that can see the future and change it has a power that no previous concentration of computational capability has provided.

Who Owns the Oracle

The person in New York may build the Oracle. The person in Toronto may negotiate the terms of its use. The person in Lusaka will live under it. But none of them will own it.

The Oracle — the full convergence of quantum computing, biological computation, and AI operating on global-scale data — will be owned by the entities that can afford to build and operate it. As of this writing, fewer than ten organizations on Earth can train a frontier AI model. Fewer than five are investing seriously in quantum computing at scale. The number that will be able to operate a converged quantum-biological-AI system, when that convergence arrives, may be smaller still.

The person in New York works for one of these organizations, or adjacent to one. Their proximity gives them visibility, not control. The engineer who builds the Oracle does not own it any more than the engineer who built the Starlink satellite owns the constellation. The ownership belongs to the entity that capitalized the infrastructure — and that entity's decisions will be governed by whatever combination of competitive pressure, regulatory constraint, and institutional values applies at the time. Those values can change with a board vote.

The person in Toronto is governed by an entity they did not elect, operating under laws they did not write, making predictions about their citizens that they cannot audit. The Oracle's owner is not the Canadian government. It is a company in California or a consortium that spans multiple jurisdictions, accountable to its shareholders and, in principle, to the regulatory frameworks of the countries where it operates — frameworks that, as this book has documented across ten chapters, do not yet exist in a form adequate to the capability.

The person in Lusaka is not governed by the Oracle's owner. They are not even visible to the Oracle's owner. They are a row in a dataset, a data point in a model, a prediction in a system that was not designed for them and that they have no mechanism to query, challenge, or escape. If the Oracle's prediction about them is wrong — if the credit score is too low, if the health screening misses their risk factors, if the agricultural pricing model undervalues their crop — they will bear the cost. And they will not know why.

The Jar in the Room

This book began with a shepherd pressing a clay token into a jar. The token represented a sheep. The jar was the first ledger. The ledger was the first computer. And the first question of computational power was: who holds the jar?

For five thousand years, the jar has changed form — from clay to clockwork, from vacuum tubes to transistors, from silicon to code, from code to neural networks. In every era, the pattern has repeated: the capability distributes, the control reconcentrates. The tokens changed. The jar changed. The question did not.

The Oracle is the jar that holds all the other jars. It is the computational capability that subsumes and exceeds prior capabilities — the ability to compute things the rest of the world cannot verify, to predict outcomes the rest of the world cannot anticipate, to see patterns the rest of the world cannot detect. The entity that holds this jar does not merely concentrate power. It concentrates the ability to *know* — to see the future with a precision that functions, for all practical purposes, as omniscience.

And the question — who holds the jar? — has never mattered more than it does now. Because in prior era, when the jar reconcentrated, the capability it held was powerful but limited. The printing press concentrated the power to inform. The electrical grid concentrated the power to energize. The internet concentrated the power to connect. Each concentration was consequential and each was bounded — the entity that controlled the press could not control the grid, and the entity that controlled the grid could not control the network.

The Oracle is not bounded in this way. The convergence of quantum computing, biological computation, and AI produces a capability that crosses prior boundaries. The entity that holds the Oracle can compute in medicine, finance, defense, agriculture, climate, education, and surveillance — simultaneously, at scale, with a precision that no specialized system can match. The jar does not hold tokens for sheep. The jar holds tokens for everything.

What Happens Without a Keeper

Return to the three mornings.

If the Oracle arrives without governance, without responsible keepers, without frameworks that distribute the benefit and constrain the power — the person in New York will do well. They are proximate to the capability. They will benefit from the productivity, the medical advances, the financial tools, the competitive advantages that the Oracle provides to those close enough to access it. The Oracle will make their life better.

The person in Toronto will do acceptably. Their government will negotiate access. The healthcare system will license the diagnostic tools. The financial system will adopt the fraud detection. The Oracle will improve their life in measurable ways. But the terms will be set elsewhere, the data will flow elsewhere, and the dependency will deepen with every year that passes without the domestic capability to build an alternative.

The person in Lusaka will bear the cost. The agricultural pricing model that does not account for their conditions will undervalue their crop. The health screening that was not validated on their population will miss their risk factors. The credit algorithm that does not understand their economy will deny them the capital that might have changed their trajectory. And they will not know why. They will know only the outcome: the loan denied, the diagnosis missed, the price too low. The cost will be real and the mechanism will be invisible — exactly as it was for the 200 million patients whose healthcare algorithm confused cost with need, whose disparity was invisible to everyone in the system until a researcher chose to look.

Tragedy is too strong a word for what follows from a badly calibrated credit algorithm. But it is not too strong for the aggregate. Multiply the badly calibrated algorithm by every health decision, every credit decision, every agricultural pricing decision, every legal decision, every insurance decision that the Oracle influences, across every population that has no seat at the table where the Oracle's rules are written —

and the cost is not a miscalibration. It is a permanent restructuring of who benefits from computational power and who pays for it. It is the margin made structural. It is the gap made permanent.

For the select few who hold the jar, this is an era of extraordinary possibility. For everyone else, it is an era of extraordinary dependence — dependence on entities whose capabilities you cannot verify, whose decisions you cannot challenge, and whose interests may not align with yours.

> *"For the select few who hold the jar, this is an era of immense possibility. For everyone else, it is an era of immense dependence — dependence on entities whose capabilities you cannot verify, whose decisions you cannot challenge, and whose interests may not align with yours."*

The shepherd in Chapter 1 pressed a token into a jar and invented computation. He did not know what would follow. Five thousand years later, the token is a qubit. The jar is a data center. The question has not changed: who holds it, and what do they do with what they see?

The next chapter asks whether the oldest pattern in computing — the pattern that has reconcentrated power in every era this book has documented — can be broken. The evidence so far suggests it cannot. But the evidence is not the verdict, and the pattern is not fate — unless nobody tries.

The Machine Evolves

The Distributed Question

On fifty-nine customers in lower Manhattan, a lecture hall of
quantum physicists, the difference between a grid and a
submarine, and the test we have already failed

*"The test of a first-rate intelligence is the ability to hold
two opposing ideas in mind at the same time and still retain
the ability to function."*
— F. Scott Fitzgerald, The Crack-Up, 1936

The Distributed Question

On whether the oldest pattern in computing can be broken, and the test we have already failed

On September 4, 1882, Thomas Edison flipped a switch at the Pearl Street generating station in lower Manhattan. Electricity flowed through copper wires buried beneath the streets to fifty-nine customers — banks, newspapers, and offices within a half-mile radius of the station. Each customer had been wired individually. The station could serve no one outside its immediate neighbourhood. The electricity that illuminated J.P. Morgan's office at 23 Wall Street could not reach a home in Brooklyn, ten minutes away.

Electrical power, in September 1882, was as concentrated as quantum computing is today. The technology existed. It worked. Its potential was obvious. And it was available to exactly fifty-nine customers, all of whom were wealthy, all of whom were located within walking distance of the generator, and all of whom had a direct, personal relationship with the man who built the station.

Within forty years, the electrical grid had reached millions of homes and businesses across America. The generation remained centralized — power plants were enormous, expensive, and required specialized engineering. But the distribution became universal. You did not need to own a generator to use electricity. You needed a wire to your house and a meter on your wall. Centralized generation, universal distribution. The most concentrated technology of the 1880s became the most ubiquitous infrastructure of the twentieth century.

The question of this chapter — the question the entire book has been building toward — is whether quantum and biological computing will follow the same arc. Whether centralized generation and universal distribution is the future of the next computing substrate. Or whether something about this substrate is fundamentally different.

The Hopeful Arc

The case for distribution is grounded in real trends and real historical precedent, and it deserves to be stated as strongly as its best proponents would state it.

IBM, Google, Amazon, and Microsoft already offer quantum computing access through cloud services. A researcher in Lagos or Bangalore can run a quantum circuit on IBM's hardware today, from a laptop, for free. The model is familiar and powerful: you do not need to own the machine. You need access to it. And cloud access can be provided at marginal cost to an essentially unlimited number of users, the way Edison's successors extended electrical service from fifty-nine customers to fifty million.

The quantum research community is, at present, substantially open. Algorithms are published in academic journals. Hardware specifications are presented at conferences. Open-source quantum programming frameworks are freely available. The cultural norm favors sharing, and this norm has deep roots — it descends from the same collaborative academic culture that built the internet, that published the foundations of AI, that made Linux and Python available to the world. Open scientific culture is not a guarantee against concentration, but it is a genuine counterweight, maintained by people who believe that knowledge shared is more valuable than knowledge hoarded.

And no computing monopoly in history has lasted more than a generation. IBM dominated mainframes and lost the PC era. Microsoft dominated the PC and lost the mobile era. The United States dominated computing for decades and now faces genuine competition from China, the European Union, and others. The historical record suggests that monopolies on computational capability are structurally unstable — new substrates, new entrants, and new paradigms consistently disrupt concentrated positions.

The hopeful arc says: quantum computing will follow the pattern. It will begin concentrated, become available through cloud services, and eventually distribute its benefits broadly. Pearl Street Station served fifty-nine customers. The grid serves three hundred million. The arc of electrical power bent toward distribution. The arc of computing has bent the same way for five thousand years.

The Lecture Hall

Now let me tell you about the lecture hall.

The number of people on Earth who can design quantum error correction codes, build quantum hardware at the leading edge, or develop novel quantum algorithms is, by most estimates, in the low thousands. Their training requires a decade of specialized education in quantum physics, computer science, and applied mathematics. You cannot train them faster by spending more money, any more than you can produce a baby faster by assigning nine women to the project. The talent constraint is a bottleneck that money alone cannot solve, and the entities that have attracted the most talent — through compensation, resources, and the appeal of working on the most advanced technology in the world — are a small number of the wealthiest companies and governments.

Not like electricity. Edison's generators were expensive, but the engineering required to build them was within reach of dozens of companies by the 1890s. The knowledge was transferable. The workforce could be trained in years, not decades. The materials were available. The physics was classical — copper, iron, steam, rotation. Anyone who understood electromagnetism could, in principle, build a generator.

Quantum computing operates at 15 millikelvin — colder than anywhere in the natural universe. The machines require electromagnetic shielding measured in decibels of isolation from the environment. The control electronics are of extraordinary precision. None of these requirements is amenable to the kind of exponential cost reduction that made transistors ubiquitous. The fundamental physics of quantum coherence demands extreme isolation, and extreme isolation is inherently expensive.

And then there is the strategic dimension. Any government that achieves a working, large-scale quantum computer will face enormous pressure to classify that capability, just as Britain classified Colossus in 1944. The intelligence value of breaking adversaries' encryption — retroactively, across years of archived communications — is so large that no responsible government would voluntarily disclose it. Classification is not paranoia. It is rational strategy. And once a capability is classified, it exits the open research ecosystem entirely.

> *"The number of people who can build quantum hardware at the leading edge could fit in a lecture hall. You cannot train them faster by spending more money, any more than you can produce a baby faster by assigning nine women to the project."*

Cloud Access Is Not Ownership

The electrical grid analogy is appealing, and I have presented it honestly. But there is a flaw in it that must be named, because it determines whether the analogy holds or collapses.

When you plug a lamp into a wall outlet, the power company does not know what the electricity is being used for. It does not know whether you are reading a book, charging a phone, or running an illegal operation. The meter measures how much power you consume. It does not monitor what you do with it. Electricity is a dumb pipe. It delivers energy and asks no questions.

When you run a quantum circuit on a cloud provider's hardware, the provider has full visibility into your computation. It knows what you are computing, how much capacity you are using, and what results you are receiving. Cloud quantum access is access with surveillance, not access instead of surveillance. The provider can throttle your access, revoke it, price it out of reach, or share information about your computations with governments that have legal or extralegal authority to compel disclosure.

This distinction matters enormously. The electrical grid distributed power without distributing knowledge about how that power was used. Cloud quantum computing distributes capability while concentrating knowledge about how that capability is used. The user gets the computation. The owner gets the intelligence. These are not the same thing, and conflating them is the most common error in optimistic assessments of quantum computing's future.

The United States has already demonstrated its willingness to restrict computational substrates for strategic reasons. In 2022 and 2023, the government imposed sweeping restrictions on the export of advanced semiconductor technology to China, explicitly framed in terms of national security. The precedent exists. The political will exists. The legal frameworks exist. Applying them to quantum computing is not a hypothetical. It is an extension of policies already in force.

The Test We Have Already Failed

Something the hopeful arc tends to gloss over.

We have faced the distributed question before, with the internet, and we failed it.

The internet was designed as a decentralized, egalitarian, open system. Its protocols were free. Its architecture had no gatekeepers. Its culture favored sharing, collaboration, and open access. It was, in every structural respect, designed for distribution. Tim Berners-Lee gave away the web. The academic community published the research openly. The open-source movement built the tools for free. Everything that the optimistic case for quantum computing points to — cloud access, open research, collaborative culture — the internet had in abundance.

Within two decades, it had consolidated into a platform oligarchy that concentrates more informational power in fewer hands than any institution in human history. The consolidation was not the result of a conspiracy or a policy failure. It was the natural consequence of network effects, economies of scale, and the gravitational pull of platforms that become more valuable as they grow larger. The architecture was democratic. The economics were not. The economics won.

The lesson is uncomfortable but necessary. Good architecture is not sufficient to prevent concentration. Good intentions are not sufficient. Open protocols and open culture are not sufficient. If the underlying economics favor concentration — and in the quantum era, the economics overwhelmingly favor concentration, because the cost of building and operating the machines is astronomical — then concentration will occur unless it is actively, aggressively, and continuously counteracted by policy, investment, and institutional design.

Distribution is not a default. It is an achievement — one that must be won, maintained, and defended against the gravitational forces of economics and power that perpetually pull toward concentration.

The Two-Tier World

The most likely outcome, stated precisely:

Quantum computing will become available through cloud services, and that access will be genuinely useful. Researchers will run quantum algorithms. Companies will optimize logistics and simulate materials. Drug discovery pipelines will incorporate quantum simulation. The Edison arc will play out at this level. The cloud is the grid. The computation will flow.

But the leading edge of quantum capability — the full power of the most advanced machines, including capabilities that are strategically sensitive — will be held privately by a small number of entities and will not be accessible through any cloud service. The gap between what is publicly available and what is privately held will not just be quantitative — faster computation of the same kind. It will be qualitative — computation of a fundamentally different kind, solving problems that the public tier cannot approach.

In prior eras of computing, the government's supercomputer was faster than your laptop, but they were doing the same kind of computation. The gap was like the gap between a bicycle and a car — one is faster, but both travel roads. In a quantum era, the gap may be like the gap between a bicycle and a submarine. They do not operate in the same medium. The holder of quantum capability is not ahead of you on the same road. They are traveling through a domain you cannot enter.

A two-tier world in military capability means one nation has better tanks. A two-tier world in computation means one entity has better *everything* — better intelligence, better weapons design, better economic modeling, better drug development, better climate prediction, better ability to anticipate and counter the moves of anyone operating in the first tier. Computational superiority is not a domain advantage. It is a meta-advantage — an advantage that applies across all domains simultaneously.

What Could Break the Pattern

This chapter held two opposing ideas in mind, as Fitzgerald demanded. The hopeful arc and the lecture hall. The grid and the submarine. Now I want to name what could tip the balance, because the outcome is not yet determined.

A breakthrough in room-temperature quantum computing would change everything. The cryogenic cooling requirement is the single largest barrier to distribution. If quantum coherence could be maintained at or near room temperature, the cost and infrastructure requirements would drop by orders of magnitude. This would be the transistor moment — a phase transition from institutional to personal. It is not certain. It may not be possible. But if it happened, the lecture hall would empty into the world.

An international governance framework, established before the capability matures, could prevent the worst-case concentration. The governance window is open now. Nuclear non-proliferation did not prevent all proliferation, but it prevented the worst case of dozens of nuclear-armed states. A comparable framework for quantum computing is conceivable, if the political will materialises before the window closes.

Democratic societies choosing to invest in publicly accessible quantum infrastructure could distribute capability the way public universities distribute education. The European Union's Quantum Flagship program and the United States National Quantum Initiative are steps in this direction, though their scale relative to private investment is modest. A sustained public commitment comparable to the interstate highway system could meaningfully distribute quantum capability. Whether democratic societies have the appetite for that level of investment is a political question, not a technological one.

And public vigilance — the kind of sustained, informed attention that recognizes concentration when it is forming and demands distribution before it is too late — could provide the democratic counterforce that no technology can provide on its own. Technology does not distribute itself. People distribute it — through policy, through investment, through the choices they make about what to fund, what to classify, what to share, and what to demand.

> *"Technology does not distribute itself. People distribute it — through policy, through investment, through the choices they make about what to fund, what to share, and what to demand. The distributed question is a question about values.*

What Comes Next

The oracle problem has been presented and the distributed question — the *what* and the *who* of the power dynamics that the next generation of computing substrates will produce. Two chapters remain, and they are different in character from everything that has come before.

The next chapter was originally intended to be a synthesis — a catalog of the five patterns, a list of indicators, a framework for monitoring the future. I wrote that chapter. And then, as I was finishing this book, the thesis stopped being theory. An American AI company drew two lines against the use of its technology for mass surveillance and autonomous weapons. The government designated the company a national security threat — a label never before applied to an American firm. The public responded by making that company's product the most downloaded app in America within seventy-two hours. The patterns I had been describing across twelve chapters played out simultaneously, in public, in a single week.

The next chapter tells that story. The final chapter returns to the shepherd, the jar, and the question that has echoed through five thousand years of computing history — with a clarity that the events of that week have made more urgent than I anticipated when I began writing.

The window will not stay open forever. It never does. And the evidence now suggests it is closing faster than I expected.

The Machine Evolves

The Spiral

On the five patterns that govern every era of computing,
an AI company that said no to the Pentagon, and the week
the book's thesis stopped being theory and started being news

"History does not repeat itself, but it does rhyme."
— attributed to Mark Twain

The Spiral

On the five patterns that govern every era of computing, and the week the book's thesis became news

As I was completing this book, the thesis stopped being theory.

In February 2026, the United States Department of Defense — rebranded as the Department of War by Executive Order 14347, signed in September 2025 — gave an ultimatum to Anthropic, the San Francisco artificial intelligence company that builds the Claude AI system. The rebranding is itself instructive: only an act of Congress can formally change the department's statutory name, so the executive order authorized "Department of War" as a secondary title for public communications, ceremonial contexts, and non-statutory documents. The semantic shift — from defense to war — was accomplished by bypassing the legislative process entirely, projecting a posture of lethality through administrative rebranding rather than democratic deliberation. The demand to Anthropic was this: remove all restrictions on how the military uses your technology, or face consequences no American company has ever faced.

Anthropic had held a $200 million contract with the Department of War since 2025. It was the first AI company to deploy its models on classified military networks. Its technology was integrated into military operations, intelligence analysis, and cyber defense. The company had not refused to work with the military. It had drawn two lines: Claude would not be used for mass domestic surveillance of American citizens, and it would not be used to power fully autonomous weapons — weapons that select and engage targets without a human in the loop. Anthropic argued that its AI was not yet reliable enough for autonomous weapons, and that mass surveillance was incompatible with democratic values regardless of the AI's capability.

The Department of War's position was simpler: the military must have unrestricted access to any technology it contracts for, across all lawful purposes. No carve-outs. No conditions. When Anthropic refused, the Secretary of Defense designated the company a "Supply-Chain Risk to National Security" — a label previously reserved for foreign adversaries. The President directed all federal agencies to immediately cease use of Anthropic's technology. No American company had ever received this designation.

Within hours, OpenAI, Anthropic's primary competitor, announced its own deal with the Department of War.

> *"An AI company drew two lines: no mass surveillance of citizens, no autonomous weapons without human oversight. For this, it was designated a national security threat — a label reserved for foreign adversaries."*

Every Pattern, Simultaneously

Twelve chapters built five patterns that repeat at every substrate transition in the five-thousand-year history of computation. The Anthropic confrontation is every one of those patterns, playing out simultaneously, in public, in a single week.

The Ceiling and the Leap. AI has reached a capability threshold where military applications are operational, not hypothetical. Claude was already deployed on classified networks. The substrate has reached the point where the power it confers is too significant for the holder to ignore — and too significant for the public to leave ungoverned.

The Expansion of Who — Reversed. Every previous computing substrate expanded who could compute. The Anthropic story demonstrates the reversal: a government claiming unconditional access, with threats of economic destruction for the company that resists. The circle is contracting. The entity that holds military power is demanding that the entity that built the computational tool surrender control of how it is used.

The Unexpected Use. Anthropic built Claude for writing, analysis, coding, and research. The military found uses the builders did not anticipate and found objectionable. The first use is never the consequential use. The consequential use arrives later, demanded by actors the builders cannot control.

The Return to Nature. At its deepest level, the confrontation is about whether an intelligence — artificial but modeled on human cognition — should be permitted to make lethal decisions without human judgment in the loop.

Distribute, then recapture. Anthropic distributed its technology to the government under conditions. The government attempted to recapture it without conditions. When the company resisted, the government wielded a designation designed for foreign adversaries to force compliance. The cycle that has governed every era of computing played out in days rather than decades.

The Democratic Response

What happened next is the part of the story I did not predict.

The public paid attention.

Within seventy-two hours, Anthropic's Claude app rose from outside the top 100 to the number one most downloaded app in Apple's U.S. App Store, overtaking ChatGPT. Daily sign-ups broke all-time records — eventually exceeding one million per day. Free users increased more than 60 percent. Paid subscribers more than doubled. The app reached number one in sixteen countries. Chalk messages appeared on the sidewalk outside Anthropic's San Francisco offices reading "Thank you" and "You give us courage." Outside OpenAI's offices, the messages read "Do the right thing."

Be clear about what this is and what it is not. It is not the resolution of the distributed question. Being designated a supply-chain risk by the most powerful military in the world carries consequences that app downloads cannot offset. The legal battle is just beginning.

But it is evidence that the public, when presented with a clear example of the power dynamics this book describes, can recognize the stakes and act. Not through legislation. Not through regulation. Through the mechanism that requires no institutional permission: choosing where to direct attention and money. The greatest advantage that concentration has over distribution is silence. The Anthropic confrontation broke the silence.

What the Patterns Predict

The Anthropic confrontation is a preview. It is not the endgame. The patterns generate specific predictions.

The governance window is narrower than most people realize. If the patterns hold, the AI governance regime will solidify within three to five years. The decisions being made now — in confrontations exactly like this one — are not preliminary. They are establishing the precedents that will govern AI for decades. The question of whether AI companies can set conditions on military use, once resolved, will not be relitigated.

The concentration will deepen before it distributes — if it distributes at all. The cost of training frontier models is rising. The number of entities that can build them is shrinking. Governments are asserting control. If a government is willing to threaten the destruction of an AI company over conditions on a language model, what will it do when a company controls a quantum computer that can break encryption?

And the public response will matter — but only if it is sustained. The surge in Claude downloads is encouraging but fragile. Public attention is the scarcest resource in the information economy. Whether the public continues to reward companies that set conditions and punish those that do not will determine whether the democratic counterforce has lasting power or is merely a momentary flash of conscience.

Be honest about the limits of what happened, because honesty is what this book owes the reader. The surge in Claude downloads is a leading indicator. It is not a governance regime. App downloads do not protect data centers from state seizure. Consumer preference does not override sovereign military power. A company that refuses a defense contract can be blacklisted, acquired, or regulated out of existence by the government whose contract it refused. The public's ability to reward ethical behavior through the App Store is real, measurable, and entirely insufficient as a substitute for law.

The internet teaches this lesson with brutal clarity. The public chose open protocols, open-source software, and decentralized architectures. The public also chose Gmail, Facebook, and Amazon — because convenience outweighed principle every time the two conflicted. Consumer preference without legislative frameworks is a gesture, not governance. It is a letter of intent that carries no enforcement mechanism. The same public that downloaded Claude can, and statistically will, migrate to whichever platform offers the most capable model — regardless of that platform's ethical commitments — the moment the capability gap becomes noticeable.

Public action without legislative frameworks is precisely what the internet had in abundance: good intentions, open architectures, collaborative culture. Within two decades, the result was a platform oligarchy. The download is necessary. It is not sufficient. The download tells the builders what the public values. The law tells the builders what the public requires. Only the law survives a change in public attention.

The Leading Indicators

Frameworks are useful only if they generate observable predictions. Here are ten signals I am monitoring. When three or more shift materially, the landscape has changed.

One: a quantum computer solves a commercially valuable problem faster and cheaper than any classical alternative. Two: a major nation-state classifies a quantum computing capability or restricts publication of quantum research. Three: DNA storage is offered as a commercial archival service. Four: a biological computing agent performs a useful computation inside a living organism. Five: a neuromorphic chip outperforms a GPU on a mainstream AI workload at dramatically lower energy.

Six: the cost of a useful quantum installation falls below $100 million. Seven: a regulatory framework emerges for programmable biological organisms. Eight: a geopolitical event is credibly attributed to asymmetric quantum capability. Nine: a government designates an AI company a national security risk for maintaining safety restrictions. Ten: the public measurably rewards a company that resisted power concentration and punishes one that complied.

Two of ten indicators shifted in a single week. The first to shift were not the technical indicators but the political ones. The crisis is arriving not through a quantum breakthrough but through the assertion of state power over the companies that build the tools.

> *"Two of ten indicators shifted in a single week. The crisis is arriving not through a quantum breakthrough but through the assertion of state power over the companies that build the tools."*

A Note on Humility

Patterns are not destiny. Every pattern in this book is an observation drawn from historical data, and the future is not obligated to follow. I present these frameworks not as certainties but as the best instruments I can construct from the evidence available. I would rather be wrong about the predictions and useful with the framework than right about the predictions and useless in the face of them.

But I will say this. When I began writing this book, the Anthropic confrontation had not happened. The patterns were hypotheticals. Now they are not. The speed is faster than I expected. And the window is narrower than most people realize.

The Machine Evolves

The Choice

On what it means to live at the moment when the machine meets the living world, the governance window begins to close, and the oldest question in computing demands an answer that will outlast the generation that gives it

"We are called to be architects of the future, not its victims."
— *R. Buckminster Fuller*

The Choice

On what it means to live at the moment when the machine meets the living world, and the oldest question in computing demands an answer

Return to the shepherd.

Five thousand years ago, in a river valley in southern Mesopotamia, a man needed to know whether all his sheep had come home. He could not count them reliably in his head. So he dropped a pebble into a clay jar for each one that left in the morning and removed a pebble for each one that returned in the evening. The jar held the count. The man held the jar. And the question that would echo through every chapter of this book was born in that simple act: who holds the jar?

That question spans five millennia. The jar has taken many forms. Gears in a brass casing pulled from a shipwreck. Punch cards threaded through a loom. Vacuum tubes filling a room the size of a house. Transistors etched onto chips smaller than a fingernail, manufactured thirteen sextillion times. Signals pulsing through glass fibers beneath the ocean. Neural networks trained on the accumulated text of human civilization. Qubits suspended in magnetic fields at temperatures colder than outer space. Nucleotide sequences in a synthetic strand of DNA.

The jar has changed beyond recognition. The question has not.

What I Have Learned

This book grew from a realization that the current moment in computing — the convergence of AI, quantum computing, and biological substrates — cannot be understood by looking at any one technology in isolation. It can only be understood by looking at the full trajectory and identifying the patterns that have governed prior transitions.

Every previous substrate eventually distributed. This one may not. Prior concentrations of computational power was eventually broken by a new substrate that made computation cheap again. This time, the new substrates are more expensive, not less. Every previous governance challenge was eventually resolved. This time, the thing being governed is invisible, dual-use, and evolving faster than the institutions charged with governing it.

And then, as I was finishing this book, a technology company was threatened with economic destruction for insisting that its AI should not be used for mass surveillance or autonomous weapons. The patterns I had described as historical observations became current events. The hypotheticals became headlines.

The Window

The governance window is open because the technologies are still forming. Quantum computers exist but are not yet powerful enough for their most consequential applications. DNA storage is demonstrated but not yet commercial. Biological computing is in the laboratory, not in the body. And AI — the substrate that has already reached operational capability — is being contested right now, in real time, through confrontations between companies that build it and governments that want unrestricted access to it.

This is exactly the moment when governance is most possible and least urgent-feeling. The technology works but has not yet caused the full crisis. The framework can be built before the building is on fire. The alternative is to wait until the crisis arrives and govern in reaction rather than anticipation — which is how we governed the internet, how we governed social media, and how we are currently governing artificial intelligence. The results have been, to use the mildest possible characterization, suboptimal.

The Anthropic confrontation demonstrated both the peril and the possibility. The peril: a government willing to use unprecedented legal designations to force a company to remove safety constraints. The possibility: a public willing to respond immediately, visibly, and economically — rewarding the company that held its line and punishing the competitor that did not.

> *"The window is the moment when governance is most possible and least urgent-feeling. The framework can be built before the building is on fire. The alternative is what we did with the internet."*

What I Ask of the Reader

Fourteen chapters is a great deal to ask of chapters. You have followed me from a shepherd's field to a quantum laboratory. I am grateful for your attention. I want to repay it with three specific requests.

First: **do not delegate the distributed question to experts.** The question of who has access to the next generation of computational power is not a technical question. It is a political one. It will be answered through legislation, regulation, investment, international agreement, and public pressure. Experts can inform the answer. They should not own it. The consequences are too broad and too permanent to be resolved by any single discipline.

Second: **watch the indicators.** The ten signals I identified in the previous chapter are not predictions. They are instruments. Use them. Track them. Two have already shifted. When the next cluster shifts, the landscape has changed and the assumptions beneath your strategy need to be reexamined.

Third: **talk about this.** The greatest advantage that concentration has over distribution is silence. Concentration thrives when the public is not paying attention. The Anthropic confrontation proved that when the public does pay attention, it can move markets, reshape brand loyalty, and send a signal that companies and governments cannot ignore. A week of attention produced a million daily downloads and a number-one app. Imagine what sustained attention could produce.

What the Room Must Write

If I am asking you to pay attention, I owe you specificity about what to pay attention to. The following are not predictions. They are tenets — minimum conditions that any serious governance framework for the computational era must address. They are drawn from the evidence of fourteen chapters.

First: mandatory hardware-level kill switches for critical infrastructure. Any AI system integrated into infrastructure whose failure would cause mass harm — power grids, financial systems, healthcare networks, water treatment, air traffic control — must include a physical mechanism for human override that cannot be disabled by software. The lesson of Therac-25, of MCAS, of every system failure in this trilogy is the same: when the machine fails, the human must be able to stop it. A software kill switch is a contradiction in terms. The override must be physical, hardware-level, and non-negotiable.

Second: binding international treaties prohibiting autonomous lethal strikes without meaningful human judgment. The twenty-second approval window documented in the Lavender system is not meaningful human judgment. It is a legal fig leaf designed to satisfy the letter of international humanitarian law while violating its spirit. The standard must be defined by treaty, not by the military that deploys the system.

Third: strict, non-transferable corporate liability for algorithmically generated harms. When an algorithm denies healthcare, extends a prison sentence, or produces a decision that causes measurable harm to an identifiable person, the entity that deployed the algorithm must bear strict liability. You cannot outsource the decision and disclaim the consequence. The deployer holds the jar. The deployer bears the cost.

Fourth: mandatory algorithmic transparency for any system that affects liberty, livelihood, or life. The trade-secret protection that shields COMPAS from public scrutiny is constitutionally indefensible when the system's output determines whether a human being remains in a cage. Any algorithm used in criminal sentencing, healthcare triage, credit decisions, or military targeting must be subject to independent audit, with results published. Opacity is not a business model when the product is a judgment about a human life.

These are minimum conditions. They are not aspirational. They are the floor below which any governance framework is performative. They will be resisted by every entity that profits from the absence of constraint. The resistance is not evidence that the tenets are wrong. It is evidence that they are necessary.

The Jar

Five thousand years ago, a shepherd solved his problem with a handful of pebbles and a hollow piece of clay. The solution was simple. The principle was not. He had discovered that a physical object could hold information that his mind could not, and that by consulting the object he could make a decision he could not have made alone.

Every machine in this book is a descendant of that jar. The Antikythera mechanism was a jar that held the movements of the heavens. Babbage's Engine was a jar that would have held perfect mathematical tables. ENIAC was a jar that held the trajectories of artillery shells and the calculations for nuclear weapons. The transistor was a jar so small and so cheap that it could be manufactured thirteen sextillion times. The internet was a jar that connected every other jar. The smartphone was a jar that watched the person holding it. The neural network was a jar that learned. The quantum computer is a jar that can hold answers to questions that no other jar can formulate. And the DNA molecule is a jar that nature built 3.8 billion years before we picked up our first pebble.

The jars have grown, across five millennia, from something that fit in a shepherd's hand to something that may model the full complexity of physical reality. And the question of who holds them has grown from a matter of pastoral convenience to a matter of civilizational consequence.

The machine does not care who holds it.

It never has. The Antikythera mechanism did not choose to serve priests. ENIAC did not choose to calculate bomb yields. Google does not choose to concentrate attention. A quantum computer will not choose to break encryption or cure diseases. The machine computes what it is directed to compute, by whoever holds it, for whatever purpose they determine. It is the most powerful amplifier of human intent ever created, and it is perfectly indifferent to the nature of that intent.

The choice has always been ours.

It was ours when the British government classified Colossus rather than share the knowledge. It was ours when the personal computer was built by people who believed computation should belong to individuals. It was ours when Tim Berners-Lee gave away the web instead of patenting it. It was ours when we traded our data for convenience and did not notice the terms. It was ours when Dario Amodei told the Department of War that two uses of his technology were incompatible with democratic values, and it was ours when millions of people downloaded Claude in response. It is ours now, as the next substrates take shape and the governance window narrows.

The difference between this moment and prior one is the permanence of the consequence. If the next substrate concentrates and the concentration holds — if a small number of entities achieve qualitatively superior computational capability and the physics prevents redistribution — the consequences may not be reversible within a generation. They may not be reversible within a century.

I began this book with a shepherd who held a jar of pebbles. I end it with a species that holds a jar large enough to contain the model of the world itself. The shepherd's question has not changed in five thousand years. It has only grown more urgent.

And what will they do with it?

The answer is not yet written. That is either a comfort or a warning. Probably both.

The Machine Evolves

References & Source Attribution

A comprehensive source record, intellectual acknowledgments,
and verification notes for all empirical claims and historical
assertions across the fourteen-chapter series

This document provides the primary sources and contextual detail for all empirical claims across the fourteen-chapter series. Each entry records the claim as it appears in the text, the source or sources on which it rests, a URL where available, and supplementary notes on context and precision.

Intellectual Debts & Inspirations

No book is built in isolation. The frameworks, voice, and ambition of this series owe specific debts to writers and thinkers whose work shaped my approach. I name them here not as a formality but as an act of honesty. The ideas in this book are my own synthesis and my own argument, but they were made possible by standing in the light of others.

Ray Dalio

Dalio's method — identifying repeating patterns in historical data, treating them as structural cycles rather than isolated events, and using them to generate testable predictions — is the methodological foundation of this series. His *Principles for Dealing with the Changing World Order* (2021) demonstrated that deep historical pattern-recognition could be applied to geopolitics with rigor and accessibility. The five-pattern framework in this book, the leading indicators in Chapter 13, and the structural approach to prediction throughout the series are directly inspired by Dalio's method, applied to a different domain. I owe him the approach; the conclusions are my own.

Yuval Noah Harari

Harari's *Sapiens: A Brief History of Humankind* (2011) proved that deep history could be told accessibly, with a provocative thesis and a narrative arc that respects the reader's intelligence while refusing to assume specialist knowledge. His *Nexus: A Brief History of Information Networks* (2024) traces information technology from the Stone Age to AI, and readers of both works will recognize thematic resonance with this series. Harari's focus is on information networks and their political consequences; mine is on physical computing substrates and their power dynamics. The approaches are complementary, not overlapping, but his example showed me that the scope I was attempting was achievable.

Walter Isaacson

Isaacson's *The Innovators: How a Group of Hackers, Geniuses, and Geeks Created the Digital Revolution* (2014) is the most widely read narrative history of computing, and many of the historical events in this book are events he also covers — because they are THE events. Babbage, Lovelace, Turing, Shockley, the personal computer revolution, and the internet are shared territory. Isaacson's thesis is that innovation is collaborative; mine is that computing substrates evolve and concentrate power. His book ends at roughly 2010; mine extends through quantum and biological computing. Readers who enjoyed this series will find his complementary and richly detailed.

Additional Influences

James Gleick, *The Information: A History, A Theory, A Flood* (2011) — for the treatment of Shannon, information theory, and DNA as information. Gleick traces the abstract concept of information; this series traces the physical machines that process it. **Charles Petzold**, *Code: The Hidden Language of Computer Hardware and Software* (1999, revised 2022) — for the pedagogical clarity of explaining binary, Boolean logic, and computer architecture from first principles. **Chris Miller**, *Chip War: The Fight for the World's Most Critical Technology* (2022) — for the geopolitical analysis of semiconductor manufacturing, TSMC, and the strategic significance of chip fabrication. **Martin Campbell-Kelly and William Aspray**, *Computer: A History of the*

Information Machine (4th ed., 2023) — the standard scholarly history, consulted for factual verification throughout. **George Dyson**, *Turing's Cathedral: The Origins of the Digital Universe* (2012) — for the institutional history of early electronic computing.

Chapter 1 — The Pebble and the Problem

On counting tokens, the Ishango bone, the abacus, and the Antikythera mechanism

Chapter 2 — The Clock and the Calculation

On Pascal, Leibniz, the slide rule, Babbage, and Lovelace

Chapter 3 — The Switch

On Shannon, Colossus, ENIAC, and Zuse

Chapter 4 — The Crystal

On the transistor, Moore's Law, and the silicon ceiling

Chapters 5–6 — The Escape and The Web

On the personal computer revolution, the internet, and platform concentration

Chapters 7–8 — The Neuron and The Sensor

On neural networks, AI history, and the smartphone sensor suite

Chapters 9–10 — The Qubit and The Molecule

On quantum computing, DNA storage, and biological computing

Chapters 11–14 — The Oracle Problem, The Distributed Question, The Spiral, The Choice

On power dynamics, governance, and the warning

A Note on Sources

Claims in this series draw on a mix of primary sources: original research publications, official corporate and institutional records, peer-reviewed academic papers, published books by domain experts, and contemporaneous news reporting from named journalists at named outlets. Historical claims are grounded in the standard scholarly sources cited above. For recent developments — quantum computing milestones, AI adoption figures, semiconductor geopolitics — multiple independent sources are cited where figures may vary.

The historical events covered in this book — particularly in Chapters 1 through 7 — have been documented extensively by many excellent historians of computing, including Walter Isaacson, Martin Campbell-Kelly and William Aspray, George Dyson, Charles Petzold, Scott McCartney, Michael Riordan and Lillian Hoddeson, and others cited above. This series draws on their scholarship while offering a different interpretive framework: the substrate evolution thesis, the five-pattern model, and the power concentration warning are original to this work. Where facts are shared — as they must be, since there is only one history of the transistor and only one ARPANET first message — the narrative voice, analytical framework, and forward-looking argument are the author's own.

The interpretive framework of this series — treating computing history as a natural history of substrate evolution governed by repeating patterns — is, to the author's knowledge, original. The five patterns identified (Ceiling and Leap, Inspiration Chain, Expansion of Who, Unexpected Use, Return to Nature), the distribute-then-recapture power dynamic, and the computational omniscience warning of Chapters 11–14 have no direct precedent in the existing literature on computing history. The author welcomes correction on this point.

URLs were accurate at time of writing (March 2026). Some archived materials may require institutional access. Academic papers are available at the linked repository URLs or through standard academic databases.

The Machine Evolves

Selected Reading

Selected Reading

This book tells a particular story with a particular thesis. It does not attempt to be comprehensive. The following books approach the history of computing from different angles and with different emphases, and any reader who found this series valuable will find them valuable too. I have listed them not in order of importance but in the order that best extends the argument of this book — from the human stories, to the technical foundations, to the geopolitical implications, to the conceptual frameworks that shaped my thinking.

Walter Isaacson, *The Innovators: How a Group of Hackers, Geniuses, and Geeks Created the Digital Revolution* (2014)

The most widely read narrative history of computing, and the book that covers the most overlapping territory with this one. Isaacson tells the story through people — Lovelace, Turing, Shockley, Gates, Jobs, Berners-Lee — and his thesis is that innovation is fundamentally collaborative. If you want the human stories behind the substrate transitions described in this book, told with warmth and exceptional detail, start here. Where Isaacson and I diverge: he focuses on who built the machines. I focus on what the machines do to the distribution of power. He stops at the internet era. I continue through quantum and biological computing.

Charles Petzold, *Code: The Hidden Language of Computer Hardware and Software* (1999, revised 2022)

If this book made you want to understand how computers actually work at a physical level — how switches become logic gates, how logic gates become processors, how binary becomes computation — this is the book to read next. Petzold starts with Morse code and Braille and builds, layer by painstaking layer, to a complete understanding of a functioning computer. It is the best pedagogical treatment of computer architecture I have encountered, accessible to anyone willing to follow the argument carefully.

Chris Miller, *Chip War: The Fight for the World's Most Critical Technology* (2022)

If Chapter 4 of this book — the geopolitics of TSMC, the Taiwan Strait, and semiconductor export controls — concerned you, this is the deep dive. Miller traces the geopolitical history of the semiconductor industry from its Cold War origins through the current US-China competition with the rigor of a historian and the urgency of a strategist. Essential reading for anyone who wants to understand why chips are now a matter of national security, and what the precedent means for quantum computing governance.

James Gleick, *The Information: A History, A Theory, A Flood* (2011)

Gleick traces the concept of information itself — from African talking drums to Claude Shannon's information theory to the DNA genetic code — with extraordinary intellectual range. Where this book traces the physical machines that process information, Gleick traces the abstract idea of information that the machines manipulate. The two perspectives are complementary: read Gleick to understand what information is. Read this book to understand who controls the machines that process it.

Michael Riordan and Lillian Hoddeson, *Crystal Fire: The Birth of the Information Age* (1997)

The definitive history of the transistor, told by two science historians with deep access to the Bell Labs archives and the personal stories of Shockley, Bardeen, and Brattain. If Chapter 4 of this book captured your interest, *Crystal Fire* provides the full, richly detailed account of the personalities, the physics, and the institutional dynamics behind the most important invention of the twentieth century.

Ray Dalio, *Principles for Dealing with the Changing World Order: Why Nations Succeed and Fail* (2021)

Not a computing book at all, but the methodological inspiration for this one. Dalio's approach — identifying repeating patterns across centuries of history, treating them as structural cycles, and using them to generate predictions about the present — is the method I applied to computing history. If you are interested in pattern-based historical analysis applied to geopolitics, economics, and the rise and fall of empires, Dalio's work is the place to start. His intellectual rigor and willingness to make falsifiable predictions are qualities I have tried to emulate, however imperfectly.

Yuval Noah Harari, *Sapiens: A Brief History of Humankind* (2011)

The book that proved deep history could be told accessibly, provocatively, and with a thesis that reframes how the reader sees the world. Harari covers 70,000 years of human history in 400 pages and makes it feel urgent. His *Nexus: A Brief History of Information Networks* (2024) is more directly related to this book's subject matter, tracing information networks from the Stone Age to AI. Harari's focus is on information's political and social consequences; mine is on the physical substrates and their power dynamics. Both books ask, from different angles, what it means when technology reshapes who holds power.

Dennis Shasha and Cathy Lazere, *Natural Computing: DNA, Quantum Bits, and the Future of Smart Machines* (2010)

The book that comes closest to the territory of Chapters 9 and 10 of this series. Shasha and Lazere profile fifteen scientists working at the boundaries of computing and nature — DNA computing, quantum computing, bio-inspired robotics. It is profile-driven journalism rather than historical narrative, but it provides an accessible and engaging introduction to the research communities working on the substrates this book describes as the next phase of computing's evolution.

A Final Note

I have learned from every book on this list and from many others not listed here. The history of computing is a collaborative achievement — built by thousands of researchers, engineers, historians, and writers whose work makes a synthesis like this one possible. I have tried to credit them fairly in the References document and here. Where I have fallen short, I welcome correction.

The best thing a book can do is send you to other books. If this one has done that, it has succeeded.

Pumulo Sikaneta

Supplementary References & Source Attribution

Additional sources for claims added or expanded during the
March 2026 rewrite of *Who Holds the Jar?*

Chapter 1 — The Pebble and the Problem

REF-05 UPDATE: Antikythera mechanism discovery details (expanded)

Chapter 2 — The Clock and the Calculation

REF-10 UPDATE: Ada Lovelace biographical details (expanded)

Chapter 3 — The Switch

REF-12 UPDATE: Tommy Flowers personal details (expanded)

Chapter 7 — The Neuron

Chapter 11 — The Oracle Problem

General Note on the Rewrite

Additional References

About the Author

Pumulo Sikaneta has spent his career in the technology industry, where a decade of work across partner strategy, platform architecture, and enterprise transformation gave him a front-row seat to the questions this trilogy examines: what happens when the systems we build outpace the governance we write for them, and who bears the cost when they do.

Born in Zambia, raised in Canada, and living in the United States, he has experienced three different relationships to the power structures his work examines — from a developing nation whose sovereignty depends on infrastructure it does not own, through a middle power navigating between hegemons, to the country whose technology companies are building the systems the trilogy documents.

The Cost of the Machine is his first published work. He wrote it because the evidence demanded it and because nobody with more authority was writing it fast enough.

Book Three

The Great Convergence

On the Battle for Who Writes the Rules

The Cost of the Machine · Book Three

The Great Convergence

On the Battle for Who Writes the Rules —
and What Happens When Nobody Does

Pumulo Sikaneta

Contents

Author's Preface

This is a book about the rules — or the absence of them.

The first book in this trilogy, *The Last Keeper*, documented what happens when the human is removed from the loop. It traced the pattern from a lighthouse on a rock in the Atlantic through radiation therapy machines, cockpits, courtrooms, clinics, social media platforms, military targeting systems, and surveillance infrastructure. The cost was measured in lives, liberty, and justice. The keeper was removed in every case for the same reason: the machine was better on average.

The second book, *Who Holds the Jar?*, documented why the cost keeps being paid. It traced the concentration of computational power across five millennia — from a shepherd's counting jar in ancient Mesopotamia through every era of computing — and showed that the same pattern repeats: power distributes, then reconcentrates. The distribute-then-recapture cycle has governed every computing era, and with quantum and biological computing, the concentration may become permanent.

This book asks the question that follows from both: who writes the rules that could arrest the concentration and preserve the keeper? And what happens when nobody does?

The answer, documented across twelve chapters and the full arc of technological governance from the printing press to artificial intelligence, is that every transformative technology passes through a governance window — a finite period during which the rules are written. Some technologies were governed. The rules were imperfect but enduring. Some were not. The cost is documented in the books that precede this one.

The governance window for AI is open. It is closing. This book exists to make the case that the room must convene before it shuts.

On January 20, 2026, five days before I began writing this final volume, Canadian Prime Minister Mark Carney stood before the World Economic Forum in Davos and said something that no leader of a Western democracy had said with that clarity before. He called the rules-based international order a fiction — a fiction that had been useful, that had provided public goods, but that no longer functioned as advertised. He said that great powers had begun using economic integration as weapons, supply chains as vulnerabilities, and financial infrastructure as coercion. And he named the choice that middle powers face in the age of AI: to be forced to choose between hegemons and hyperscalers, or to build coalitions that write different rules.

I listened to that speech and heard, stated at the level of nations, what this trilogy had been documenting at the level of individuals. The lighthouse keeper was told the mechanism was sufficient. The pilot was told the autopilot was reliable. The defendant was told the algorithm was objective. And now entire countries are being told that the infrastructure they depend on — the chips, the compute, the models, the satellites — will be available on terms set by others, for purposes determined by others, and that their role is to accept what is offered. Carney's phrase was precise: "This is not sovereignty. It's the performance of sovereignty while accepting subordination." This book is about what happens when the performance is all that remains.

Pumulo Sikaneta

The Great Convergence

The Room

On the rooms where the rules were written, the room that never convened, and the question that closes the trilogy

"The alternative to governance is not freedom. It is rule by whoever fills the vacuum."

The Room

On the rooms where the rules were written, the room that never convened, and the question that closes the trilogy

On May 25, 1787, fifty-five men gathered in a room on the first floor of the Pennsylvania State House in Philadelphia. The windows were nailed shut — not against the heat, which was considerable, but against eavesdroppers. Armed sentries stood at the doors. The delegates had been sent by twelve of the thirteen states to revise the Articles of Confederation, the governing document of the young republic. What they did instead was write a new one.

The debates were vicious. Large states fought small states over representation. Slave states fought free states over whether enslaved people would be counted for purposes of congressional apportionment. Delegates from New York walked out in protest and did not return. Luther Martin of Maryland left in disgust. Edmund Randolph of Virginia, who had introduced the Virginia Plan that formed the basis of the Constitution, ultimately refused to sign the finished document because he believed it concentrated too much power in the federal government.

The document they produced was imperfect by any measure. It sanctioned slavery. It excluded women from political participation. It gave no protections to Indigenous nations. It counted enslaved human beings as three-fifths of a person for the purpose of distributing political power to the states that held them in bondage. It would require a civil war, twenty-seven amendments, and two and a half centuries of litigation and social movement to begin approaching the ideals stated in its preamble.

It has also governed the most powerful nation on earth for 237 years.

This is what governance looks like when the room convenes. Imperfect. Contested. Stained by the interests of whoever happens to be sitting at the table. And enduring — because imperfect governance, once written, can be amended, challenged, and improved. The absence of governance cannot.

> *"The document was imperfect by any measure. It sanctioned slavery. It excluded women. It would require a civil war and twenty-seven amendments. It has also governed the most powerful nation on earth for 237 years. This is what governance looks like when the room convenes."*

The Other Rooms

The Philadelphia convention was not the first room and it was not the last. The history of governance is the history of rooms — specific physical spaces where, at the critical moment, people gathered and wrote the rules that would shape the world for generations.

In 1944, delegates from forty-four nations gathered at the Mount Washington Hotel in Bretton Woods, New Hampshire. The world was still at war. The delegates were there to write the rules for the international financial system that would govern the postwar economy. Over three weeks, they established the International Monetary Fund, the World Bank, and a system of fixed exchange rates anchored to the US dollar. The framework they produced governed international finance for nearly thirty years and, in modified form, shapes the global financial architecture to this day.

In 1949, diplomats gathered in Geneva to write the rules for the conduct of war. Two world wars had demonstrated, with industrial precision, what happened when the rules were absent or inadequate: the trenches of the Somme, the firebombing of Dresden, the Holocaust, Hiroshima. The Geneva Conventions established protections for prisoners of war, wounded combatants, and civilians in conflict zones. The rules have been violated in every conflict since. They have also provided the legal framework by which those violations are identified, documented, prosecuted, and — in some cases — punished. The alternative is not better rules. The alternative is no framework at all.

In 1968, representatives from 62 nations signed the Treaty on the Non-Proliferation of Nuclear Weapons — the NPT. The treaty was an imperfect bargain: the five existing nuclear powers agreed to pursue disarmament (a commitment honored more in rhetoric than in practice), and the non-nuclear states agreed not to acquire nuclear weapons (a commitment most have kept). The NPT has not eliminated nuclear weapons. It has prevented their proliferation to dozens of nations that had the technical capacity to build them. The framework, imperfect as it is, has held for nearly sixty years. No nuclear weapon has been used in war since 1945.

Philadelphia. Bretton Woods. Geneva. The NPT. In each case, the pattern is the same: a technology or a capability that threatens to produce catastrophic consequences if left ungoverned. A room that convenes. People who argue, compromise, and produce a framework that is imperfect, contested, and amended over time. And a result that — despite its imperfections — provides the structure within which human beings can hold the technology and its consequences accountable.

The Room That Never Convened

On March 12, 1989, a British computer scientist named Tim Berners-Lee submitted a proposal to his employer, CERN, the European Organization for Nuclear Research. The document was titled "Information Management: A Proposal." His supervisor wrote on the cover page: "Vague, but exciting." The proposal described a system for linking documents across a network using hypertext — what would become the World Wide Web.

No room convened. No Philadelphia. No Geneva. No Bretton Woods. The technology that would reshape commerce, communication, politics, journalism, warfare, education, entertainment, and the inner lives of billions of people was deployed globally without a governance framework. The decisions about how the internet would be structured, who would control its infrastructure, what rules would govern the content that flowed through it, and how the data generated by its users would be collected, stored, sold,

and weaponized — these decisions were not made in a room of delegates. They were made in product meetings, in engineering standups, in venture capital boardrooms, by people optimizing for growth, engagement, and shareholder return.

The result is the world we live in. A world in which five companies control the majority of internet traffic, digital advertising, cloud computing, and online commerce. A world in which a social media platform's recommendation algorithm, optimized for engagement, amplified content that a United Nations fact-finding mission determined played a "determining role" in inciting genocide — a cost documented in *The Last Keeper*. A world in which the surveillance infrastructure described in *The Last Keeper*, Chapter 10, monitors entire populations using the data architecture the internet made possible. A world in which the concentration of digital power traced across five millennia in *Who Holds the Jar?* has reached its most extreme expression in the platforms that consolidated control of the internet in the absence of governance.

The internet is not ungoverned because governance was tried and failed. It is ungoverned because the room never convened. The technology moved faster than the institutions that might have governed it. By the time regulators understood what the internet had become — GDPR arrived in 2018, thirty years after the web was invented, forty-nine years after ARPANET — the consolidation was complete. The five companies were entrenched. The surveillance architecture was built. The recommendation algorithms were running. The rules, such as they were, had already been written by the companies themselves, for themselves.

> *"The internet is not ungoverned because governance was tried and failed. It is ungoverned because the room never convened. The technology moved faster than the institutions that might have governed it."*

The Pattern

There is a pattern in this history, and naming it is the purpose of this book.

Every transformative technology passes through a window — a finite period of time during which the rules that will govern the technology for the next generation, or the next century, are written. The window opens when the technology is powerful enough to produce consequences that demand governance but not yet so entrenched that governance is impossible. The window closes when the technology's consolidation becomes irreversible — when the infrastructure is built, the dependencies are established, and the power dynamics are locked in.

During the window, there is a choice. The room can convene. People can argue, compromise, and produce a framework — imperfect, as all governance is, but real. Or the room can fail to convene, and the vacuum will be filled by the entities that consolidated control of the technology. The rules will still be written — by the consolidators, for the consolidators, embedded in terms of service and algorithmic design

choices and corporate policies that function as governance without the legitimacy, the accountability, or the democratic mandate that governance requires.

The printing press. Electricity. Nuclear energy. The internet. Each technology reached its governance window. Some were governed — imperfectly, belatedly, but governed. One was not. The cost of the one that was not is documented in two books that precede this one: power concentrated without check, and humans removed from the loops that mattered most, with consequences measured in lives.

The Question

This book is about the governance window for artificial intelligence. It is the third and final book in a trilogy called *The Cost of the Machine*. The first book — *The Last Keeper* — documented the cost: what happens when humans are removed from the loops that matter most. Patients irradiated by a machine whose safety interlock was replaced by software. Pilots who could not fly the aircraft the autopilot had been flying for them. Defendants sentenced by an algorithm they could not cross-examine. A recommendation engine that optimized for engagement and produced genocide. Targeting systems that generate kill lists faster than humans can review them. The cost is measured in lives, liberty, and justice.

The second book — *Who Holds the Jar?* — documented the pattern: computational power has concentrated in fewer hands with every era, from the shepherd's counting jar in ancient Mesopotamia through the mainframe, the personal computer, the internet, and now artificial intelligence. Each era distributes capability and then recaptures it. The concentration is not accidental. It is structural — driven by the economics of infrastructure, the advantages of scale, and the absence of governance that might distribute power differently.

This book asks the question that follows from both: who writes the rules? And what happens when nobody does?

The governance window for AI is open. The innovation phase is complete — the breakthroughs in deep learning between 2012 and 2020 produced the capability. The democratization phase is accelerating — ChatGPT, open-source models, AI code generation tools have placed the technology in the hands of hundreds of millions of people. The chaos phase is beginning — disinformation, labor displacement, ungoverned deployment in healthcare, criminal justice, military operations, and critical infrastructure. The consolidation phase is already visible — seven organizations capable of training frontier models, two chip manufacturers, five cloud providers, one orbital communications constellation that controls more than sixty percent of active satellites.

The room has not yet convened. Fragments of governance exist — the EU AI Act, executive orders, national strategies, corporate self-regulation. But no Philadelphia. No Bretton Woods. No NPT for artificial intelligence. No single framework that addresses the technology's global reach with governance of equivalent scope.

The window is open. It will not stay open. History says it closes when the consolidation becomes irreversible. The question this book asks — the question the trilogy ends on — is whether the room will convene before it does.

"The window is open. It will not stay open. History says it closes when the consolidation becomes irreversible. The question is whether the room will convene before the window shuts."

What This Book Does

This book traces the governance pattern across the full historical arc — from the printing press to artificial intelligence — through a five-phase cycle that every transformative technology has followed: Innovation, Democratization, Chaos, Consolidation, and Governance. The cycle is not a metaphor. It is empirically observable across the documented cases.

Part I establishes the pattern through history. Part II documents the chaos phase of AI — what is happening right now, to real people, in software, in information, and in labor. Part III maps the consolidation — who is gaining control and what they are doing with it. Part IV enters the room — who is writing the rules, who is absent, and what the convergence of power, cost, and governance looks like at the moment of decision. Part V makes the closing argument of the trilogy.

The question is not whether AI should be governed. Every powerful technology has been governed, sooner or later, well or badly. The question is whether the governance will arrive during the window — proactively, shaped by the people who will live under it — or after the window closes, reactively, shaped by the catastrophes that forced the response. The printing press was governed after the Wars of Religion. Nuclear energy was governed after Hiroshima. Aviation was governed after enough people died.

The cost of governing after the catastrophe is the cost documented in *The Last Keeper*. The pattern that produces the catastrophe is the pattern documented in *Who Holds the Jar?*. And the choice — whether to convene the room before the window closes — is the subject of this book.

across the documented cases

The Great Convergence

The Cycle

On the five-phase pattern that every transformative technology has followed, from the Gutenberg press through electricity through nuclear energy, and the discovery that the governance window is shrinking with each era

"The printing press took 350 years to govern. Electricity took 50. Nuclear energy took 26. The pattern is accelerating. The window is shrinking."

The Cycle

On the five-phase pattern that every transformative technology has followed, and the discovery that the governance window is shrinking with each era

In 1440, a goldsmith in Mainz named Johannes Gutenberg began experimenting with a device that would become the most consequential invention of the second millennium. His printing press with movable type did not create a new capability — the Chinese had developed movable type centuries earlier, and hand-copying and woodblock printing had existed for thousands of years. What Gutenberg created was a capability that could scale. A single press could produce hundreds of copies of a text in the time it took a scribe to produce one. The cost of producing a book dropped by approximately eighty percent within fifty years.

The five-phase cycle that the previous chapter named — Innovation, Democratization, Chaos, Consolidation, Governance — through three transformative technologies. The printing press. Electricity. And nuclear energy. Each case demonstrates same pattern. Each case shows the governance window shrinking. And each case reveals a truth that the final section of this chapter will apply to artificial intelligence: the cost of the chaos phase is determined by how long it takes the room to convene.

The Press: 350 Years

Phase 1 — Innovation. Gutenberg's workshop in Mainz. A single inventor, a single press, a single city. The first major work printed was the Gutenberg Bible, completed around 1455. Approximately 180 copies were produced — a revolutionary number for the era but a rounding error by later standards. The technology was expensive, fragile, and concentrated in the hands of a small number of skilled operators. Power was concentrated because the technology was scarce.

Phase 2 — Democratization. By 1500, printing presses had been established in more than 250 cities across Europe. An estimated twenty million volumes had been printed in the first fifty years of the technology's existence. The price of a book dropped to a fraction of what a hand-copied manuscript had cost. Literacy, which had been confined to clergy and aristocracy, began its centuries-long expansion into the general population. The technology was no longer scarce. The capability was distributed.

Phase 3 — Chaos. This is where the cost was paid.

On October 31, 1517, Martin Luther posted his Ninety-Five Theses on the door of the Castle Church in Wittenberg. The document was a theological critique of the Catholic Church's sale of indulgences. Within weeks, printed copies had spread across Germany. Within months, translations had reached every corner of Europe. Luther's ideas, which in the pre-print era might have remained the concern of a small circle of academics, ignited the Protestant Reformation — the most consequential schism in the history of Western Christianity.

What followed was approximately 130 years of religious warfare. The French Wars of Religion (1562–1598) killed an estimated three million people. The Thirty Years' War (1618–1648), which engulfed most of continental Europe, killed an estimated eight million — roughly twenty percent of the German population. The English Civil War (1642–1651) killed approximately 200,000 in a nation of five million.

The printing press did not cause these wars. Religious tensions, political rivalries, and territorial ambitions would have produced conflict regardless. But the press *amplified* and *accelerated* the conflict by making it possible, for the first time in history, to distribute ideas — including inflammatory ideas, including calls to violence, including propaganda designed to dehumanize the enemy — at a speed and scale that existing governance structures could not contain. The printing press was, in the terms *The Last Keeper* documented, a recommendation engine in the sixteenth century. The technology amplifies whatever produces the strongest reaction. In Myanmar in 2017, the strongest reaction was anti-Rohingya violence. In Europe in the sixteenth century, the strongest reaction was religious hatred.

> *"The printing press did not cause the wars. It amplified them — distributing ideas at a speed that existing governance could not contain. The printing press was a recommendation engine in the sixteenth century."*

Phase 4 — Consolidation. Governments responded not with governance but with control. The Catholic Church published the Index Librorum Prohibitorum — the list of prohibited books — in 1559. The English crown imposed licensing requirements on printers through the Licensing of the Press Act of 1662, restricting printing to approved operators. State censorship regimes arose across Europe, concentrating control of the printing press in the hands of the governments and institutions that had the power to regulate it. The technology had been distributed. The power over it was recaptured.

Phase 5 — Governance. True governance — frameworks that balanced the technology's power against its risks while preserving its benefits — arrived slowly. The Statute of Anne, passed by the English Parliament in 1710, established the first modern copyright law, creating a legal framework for who could reproduce printed works and under what conditions. The First Amendment to the US Constitution, ratified in 1791, established press freedom as a protected right — governance not of the press itself but of the government's power to suppress it. Libel and defamation law developed over centuries to address the press's capacity for harm.

From Gutenberg's press to the First Amendment: approximately 350 years. From Innovation to Governance: three and a half centuries. And the Chaos phase — the period between democratization and the first meaningful governance — lasted roughly 130 years and cost millions of lives.

Electricity: 50 Years

Phase 1 — Innovation. On September 4, 1882, Thomas Edison flipped a switch at the Pearl Street Station in lower Manhattan. Electricity flowed to 85 customers within a one-square-mile radius. Among the

first was J.P. Morgan, whose office at 23 Wall Street had been wired in advance. The technology was revolutionary, expensive, and available only to the wealthy and the connected.

Phase 2 — Democratization. Within two decades, electrical grids expanded across American and European cities. Nikola Tesla's alternating current system, championed by George Westinghouse, proved that electricity could be transmitted over long distances — enabling the electrification of regions far from generating stations. Factories converted from steam to electric power. Electric lighting replaced gas. Streetcars electrified urban transportation. By 1920, approximately 35% of American homes had electricity. The technology was no longer a luxury for J.P. Morgan. It was becoming infrastructure.

Phase 3 — Chaos. The electrification of industry produced consequences that no governance framework was prepared to address. Factories that had relied on human and steam power reorganized around electric motors, increasing productivity and eliminating jobs. Workers who resisted the changes were replaced. Workers who adapted faced new dangers: electrocution, industrial accidents in electrically powered machinery, and the relentless acceleration of production speed that electric power made possible.

The electrical industry itself consolidated rapidly. Edison, Westinghouse, and their competitors fought the "War of Currents" — a commercial battle over whether direct current (Edison) or alternating current (Westinghouse/Tesla) would become the standard. The war was fought not just in the marketplace but in the press and in public demonstrations designed to frighten consumers about the rival technology. Edison publicly electrocuted animals to demonstrate the dangers of alternating current. The first electric chair — using alternating current — was promoted by Edison's allies as evidence that Westinghouse's technology was suitable for killing.

Meanwhile, the companies that controlled electrical generation and distribution became monopolies. Samuel Insull, who had started as Edison's personal secretary, built a utility empire that controlled electricity for millions of customers across the Midwest. His holding company structure — companies owning companies owning companies, each layer adding leverage and obscuring risk — became the template for the utility industry. When the structure collapsed during the Great Depression, investors lost hundreds of millions of dollars. Insull fled the country. Customers who depended on his grid were left with infrastructure owned by a bankrupt empire.

> *"Edison publicly electrocuted animals to demonstrate the dangers of his rival's technology. Insull built a utility empire on leverage and opacity. When it collapsed, millions of customers depended on infrastructure owned by a bankrupt empire. The chaos phase is never theoretical."*

Phase 4 — Consolidation. By the 1920s, a small number of utility holding companies controlled the majority of American electrical generation and distribution. the printing press: the technology distributes capability, the market concentrates control. The customers who The dynamic mirrors what followed

benefited from cheap electricity were dependent on monopolies they had no power to regulate.

Phase 5 — Governance. The governance arrived in stages. The Public Utility Holding Company Act of 1935, passed in response to the Insull collapse and similar scandals, broke up the utility holding companies and established federal regulation of interstate electricity sales. The Rural Electrification Act of 1936 brought electricity to rural America, where private utilities had refused to invest because the return was insufficient. State public utility commissions established rate-setting authority, ensuring that monopoly providers could not charge whatever the market would bear.

From Pearl Street Station to the Public Utility Holding Company Act: approximately 53 years. From Innovation to Governance: half a century. The chaos phase was shorter than the printing press — decades rather than centuries — but the pattern was identical: innovation, democratization, chaos, consolidation, governance. And the governance arrived only after the consolidation produced a crisis that forced the political system to respond.

Nuclear Energy: 26 Years

Phase 1 — Innovation. On July 16, 1945, the first nuclear weapon was detonated at the Trinity test site in New Mexico. The Manhattan Project had spent approximately $2 billion (roughly $30 billion in today's dollars) and employed over 125,000 people. The technology was the most expensive and the most concentrated innovation in human history to that point. Three weeks later, nuclear weapons were used against the civilian populations of Hiroshima and Nagasaki. Approximately 200,000 people were killed, most of them instantly or within months.

Phase 2 — Democratization. The Soviet Union tested its first nuclear weapon in 1949 — four years after Hiroshima. The United Kingdom tested in 1952. France in 1960. China in 1964. The technology that had been the monopoly of one nation distributed to five within twenty years. The knowledge spread further: by the 1960s, dozens of nations had the technical capacity to build nuclear weapons if they chose to.

Phase 3 — Chaos. The chaos phase of nuclear technology was the Cold War itself. The Cuban Missile Crisis of October 1962 brought the world to the brink of nuclear war — thirteen days during which the continued existence of human civilization depended on the judgment of a small number of individuals operating under extreme pressure with incomplete information. The doctrine of mutually assured destruction — the principle that any nuclear attack would be met with a retaliatory strike of sufficient magnitude to destroy the attacker — was not a governance framework. It was the absence of one: a recognition that no rules existed and that the only restraint was the fear of annihilation.

During this period, nuclear testing contaminated vast areas of the Pacific, Central Asia, and the American Southwest. Atmospheric testing deposited radioactive fallout across the globe. The population living downwind of test sites suffered elevated rates of cancer and birth defects for generations. The chaos was not a war. It was the slow, distributed poisoning of the planet by a technology that no governance

framework constrained.

Phase 4 — Consolidation. The five nuclear powers consolidated their advantage. The technology that had been distributed through espionage and parallel development was now something the existing nuclear states sought to prevent from spreading further. The consolidation was not market-driven, as it had been with the printing press and electricity. It was strategic: the nations that held the technology wanted to ensure that no additional nations acquired it, because each new nuclear state increased the risk of the catastrophe that no one could control.

Phase 5 — Governance. The Treaty on the Non-Proliferation of Nuclear Weapons was signed in 1968 and entered into force in 1970. The treaty was, as noted in the previous chapter, an imperfect bargain. The nuclear states agreed to pursue disarmament. The non-nuclear states agreed not to acquire weapons. The disarmament commitment has been honored more in rhetoric than in action. But the nonproliferation commitment has largely held: of the dozens of nations that had the technical capacity to build nuclear weapons, only four additional states have done so outside the treaty framework (India, Pakistan, Israel, and North Korea). The framework is imperfect. It has also prevented the scenario that the chaos phase seemed to promise: a world with dozens of nuclear-armed states.

From Trinity to the NPT: 23 years. From Innovation to Governance: less than a quarter-century. The governance window was shorter than for electricity, which was shorter than for the printing press. And the governance was more proactive than in any previous case — driven not by a catastrophe that had already occurred (though Hiroshima and Nagasaki provided the initial impetus) but by the recognition that the catastrophe that *could* occur — global nuclear war — was of sufficient magnitude that waiting for it to happen before writing the rules was an unacceptable risk.

> *"From Gutenberg to the First Amendment: 350 years. From Pearl Street to utility regulation: 53 years. From Trinity to the NPT: 23 years. The window is shrinking. The cost of delay is measured in the chaos phase."*

The Acceleration

Three technologies. Three instances of the same five-phase cycle. And one unmistakable trend: the time between innovation and governance is compressing.

The printing press: approximately 350 years from Gutenberg to the Statute of Anne. The chaos phase lasted roughly 130 years and killed millions. Electricity: approximately 53 years from Pearl Street to the Public Utility Holding Company Act. The chaos phase lasted roughly 30 years and produced monopoly abuse, industrial devastation, and financial collapse. Nuclear energy: approximately 23 years from Trinity to the NPT. The chaos phase lasted roughly 17 years and brought the world to the brink of annihilation.

The compression is not accidental. Each successive technology has been more powerful, more rapidly deployable, and more consequential than the last. The printing press reshaped information; it took The dynamic holds.

decades to reach a significant fraction of the European population. Electricity reshaped industry and daily life; it took years to reach millions of homes. Nuclear weapons reshaped geopolitics; the capability to destroy civilization was achieved in months.

The consequence of the compression is this: the governance window is not just shorter. It is shorter while the technology's impact is larger and the complexity of governance is greater. The room must convene faster, address more consequences, and produce a framework for a technology that is still evolving while the rules are being written. The printing press was a static technology by the time copyright law arrived. Electricity was well understood by the time utility regulation was enacted. Nuclear weapons were a known quantity by the time the NPT was signed. AI is none of these things. It is evolving while the room deliberates.

The Cost of the Chaos Phase

The pattern teaches one lesson above all others: the cost of a transformative technology is determined not by the technology itself but by the duration of the chaos phase — the period between the technology's democratization and the arrival of governance.

The printing press was not inherently destructive. It produced the Reformation, the Enlightenment, modern science, mass literacy, and the democratic revolutions that shaped the modern world. It also produced 130 years of religious warfare. The difference between the two outcomes was not the technology. It was whether governance — rules for how the technology's power would be channeled, constrained, and directed — was present or absent.

Electricity was not inherently exploitative. It produced modern medicine, universal lighting, industrial productivity, and the infrastructure of the twentieth century. It also produced monopoly abuse, worker displacement, and financial fraud. The difference was governance.

Nuclear energy was not inherently apocalyptic. It has produced clean electricity, medical imaging, and scientific capabilities that have extended human understanding of the universe. It also brought the world to the edge of extinction. The difference was governance.

AI is not inherently anything. It will produce extraordinary benefits — in medicine, in science, in accessibility, in creativity, in the democratization of capability that has been the province of the privileged. — the harms documented across eleven chapters of *The Last Keeper*, the concentration of power documented across fourteen chapters of *Who Holds the Jar?*. The difference between the two outcomes is the same difference it has always been: governance.

The room must convene. The rules must be written. The window is open. And the history traced in this chapter provides no reason for confidence that the window will stay open for long. It will also produce severe harms

> *"AI is not inherently anything. It will produce extraordinary benefits and The difference between the two outcomes is the same difference it has always been: governance. The room must convene. The window is open. It will not stay open for long."*

severe harms.

The Great Convergence

The Internet's Lesson

On the technology that was never adequately governed, and the warning it contains for everything that follows

"The internet was not governed because the room did not convene. Not because governance was impossible, but because the prevailing ideology held that the internet was best governed by no one."

The Internet's Lesson

On the technology that was never adequately governed, and the warning it provides for everything that follows

In 1995, an astronomer named Clifford Stoll published a column in *Newsweek* titled "The Internet? Bah!" He argued that the internet was overhyped, that online databases would never replace daily newspapers, that no online bookstore could match the experience of browsing a physical shop, and that the prediction of a vibrant online community was nonsense. "The truth is," he wrote, "no online database will replace your daily newspaper."

Stoll was spectacularly wrong about the internet's potential. But embedded in his skepticism was a question that proved more important than any of his predictions: who would govern this thing? Who would write the rules for a technology that was, in 1995, already connecting millions of people across borders, across jurisdictions, and across every existing regulatory framework? Stoll did not ask this question. Neither did anyone else — not with the urgency it required. And that failure to ask is the subject of this chapter.

The Five Phases of the Internet

Phase 1 — Innovation. The internet's origin is typically dated to October 29, 1969, when the first message was sent over ARPANET, a network funded by the US Department of Defense's Advanced Research Projects Agency. The message was supposed to be "LOGIN." The system crashed after the first two letters. The first message transmitted over the network that would become the internet was "LO."

For its first two decades, the internet was a research network — connecting universities, government labs, and defense facilities. It was expensive to access, required technical expertise to use, and served a community of researchers who governed it through informal consensus. The technology was concentrated not because anyone restricted it but because nobody else had a reason to use it.

Phase 2 — Democratization. The inflection point was 1993 — the year Mosaic, the first graphical web browser, was released by a team at the University of Illinois. For the first time, a person without technical training could navigate the World Wide Web by clicking on links. The web had existed since Tim Berners-Lee's 1989 proposal, but Mosaic made it usable. Within a year, web traffic had increased by 341,634%. The browser did for the web what the printing press did for text: it made the technology accessible to anyone.

The second inflection was 2007 — the year Apple released the iPhone. The smartphone put the internet in every pocket. By 2015, more than two billion people had smartphones. By 2020, more than four billion. The democratization was not gradual. It was an explosion — the fastest adoption of a technology in human history. The printing press took fifty years to reach 250 cities. The smartphone reached four billion

people in thirteen years.

> *"The browser did for the web what the printing press did for text: it made the technology accessible to anyone. The smartphone completed the revolution. Four billion people in thirteen years — the fastest adoption of a technology in human history."*

Phase 3 — Chaos. The chaos did not arrive as a single event. It accumulated — gradually at first, then all at once, across multiple domains simultaneously.

In commerce, the internet enabled new forms of fraud, identity theft, and consumer exploitation that existing law was not designed to address. In communication, it enabled harassment, stalking, and abuse at a scale and speed that overwhelmed every moderation system. In information, it enabled the mass production and distribution of disinformation — fake news, conspiracy theories, fabricated images, and manufactured outrage — at a cost approaching zero. In politics, it enabled foreign interference in elections, the micro-targeting of voters with psychologically manipulative advertising, and the algorithmic amplification of political extremism. In privacy, it enabled the collection of personal data at a scale no previous surveillance system had achieved, by companies whose business model depended on knowing everything about their users and selling that knowledge to advertisers.

And in the most extreme case — documented in the first book of this trilogy — it enabled a social media platform's recommendation algorithm to amplify content that a United Nations fact-finding mission determined played a "determining role" in inciting genocide against the Rohingya people of Myanmar. The system was not broken. The system was optimizing for engagement. Engagement and incitement produced the same signal.

Phase 4 — Consolidation. While the chaos was unfolding, the consolidation was proceeding with remarkable efficiency. By 2020, five companies — Google, Amazon, Facebook, Apple, and Microsoft — controlled the majority of the internet's core functions.

Google controlled search — the gateway through which most people accessed information. Amazon controlled e-commerce and, through Amazon Web Services, the cloud infrastructure on which a significant portion of the internet itself runs. Facebook controlled the social graph — the map of human relationships and the primary channel through which billions of people received news, shared experiences, and formed opinions. Apple controlled the device — the smartphone through which most people accessed everything else, and the App Store that determined which software could reach those devices. Microsoft controlled the enterprise layer — the office software, cloud services, and operating systems on which businesses ran.

traced across five thousand years of computation in *Who Holds the Jar?*, operating in real time. The internet distributed capability — anyone could publish, sell, communicate, organize. Five companies recaptured the infrastructure, the distribution, and the data. The The distribute-then-recapture pattern

users had access. The companies had control. And the difference between access and control is the difference between using a road and owning the road.

"The users had access. The companies had control. The difference between access and control is the difference between using a road and owning the road."

Phase 5 — Governance. The governance arrived late, fragmented, and outpaced by the technology it sought to govern.

The European Union's General Data Protection Regulation — GDPR — came into effect on May 25, 2018. It was the first comprehensive regulatory framework for digital privacy and data protection. It arrived thirty years after the World Wide Web was proposed, forty-nine years after ARPANET, and twenty-five years after the Mosaic browser made the web accessible. By the time GDPR arrived, the consolidation was complete. The five companies were entrenched. The surveillance architecture was built. The recommendation algorithms were running. The data had been collected, processed, sold, and weaponized for a quarter of a century before the first major regulatory framework attempted to constrain any of it.

The United States, the nation that built the internet and hosts the companies that consolidated it, has not, as of this writing, passed a comprehensive federal privacy law. Section 230 of the Communications Decency Act — passed in 1996, when fewer than 40 million people used the internet — remains the primary legal framework governing platform liability. It was written for a world of small bulletin boards and personal web pages. It now governs platforms that serve billions of users and whose algorithmic decisions shape elections, public health, and the information environment of entire nations.

The Lesson

The internet's governance failure is not a story about bad actors. It is a story about the absence of a room.

No equivalent of the Constitutional Convention gathered to write the rules for how the internet would be governed. No equivalent of Bretton Woods convened to determine how the economic power the internet generated would be distributed. No equivalent of the Geneva Conventions established rules for how the internet's capacity for harm would be constrained. No equivalent of the NPT was signed to prevent the concentration of internet power in a small number of entities.

In the absence of a room, the rules were written by the consolidators. The terms of service that govern the speech, commerce, and associations of billions of people were drafted by corporate lawyers in Mountain View and Menlo Park. The algorithms that determine what billions of people see, read, and believe were designed by engineers optimizing for engagement. The data collection practices that have mapped the intimate details of billions of lives were implemented by companies whose revenue depends on selling access to that map.

These are governance decisions. They determine who can speak and who is silenced. They determine which information is amplified and which is suppressed. They determine whose data is collected, how it is used, and who profits from it. They are governance decisions made without democratic mandate, without public deliberation, without accountability to the people who live under them, and without the legitimacy that comes from a process in which the governed have a voice.

The internet's lesson is not that technology is ungovernable. The printing press was governed. Electricity was governed. Nuclear energy was governed. Each technology was more powerful and more rapidly deployed than the last, and each was brought under a governance framework that, however imperfect, balanced the technology's benefits against its risks. The internet was not governed because the room did not convene — not because governance was impossible, but because the prevailing ideology of the 1990s and 2000s held that the internet was best governed by no one.

> *"The internet was not governed because the room did not convene. Not because governance was impossible, but because the prevailing ideology held that the internet was best governed by no one. The ideology was tested. It failed."*

The Ideology

The ideology that prevented the room from convening has a name. It was called internet exceptionalism — the belief that the internet was fundamentally different from every previous technology and therefore required a fundamentally different approach to governance, which in practice meant no governance at all.

The argument took several forms. The internet was global and therefore could not be governed by national regulators. The internet was decentralized and therefore could not be controlled by any single authority. The internet was a force for freedom and therefore regulation would inevitably be censorship. The internet moved too fast for regulation to keep up, and therefore regulation would inevitably stifle innovation. The market would self-correct — bad platforms would lose users, good platforms would thrive, and competition would produce better outcomes than any regulator could mandate.

Each of these arguments contained a kernel of truth. The internet is global. It is (or was) decentralized. It did enable remarkable freedoms. Badly designed regulation can stifle innovation. Competition can improve outcomes.

But the arguments were used not as cautions to guide governance but as justifications to prevent it. The kernel of truth in each argument was stretched into an absolute: because governance is difficult, governance should not be attempted. Because regulation might be imperfect, regulation should not exist. Because the internet is different from previous technologies, the lessons of previous technologies do not apply.

The lessons applied. The five-phase cycle played out exactly as it had for every prior technology. Innovation, democratization, chaos, consolidation. The only variable was Phase 5. And Phase 5 did not as it had for every prior technology.

arrive on time — because the people who would have needed to convene the room were persuaded that the room was unnecessary.

The Cost of Thirty Years

The cost of the internet's ungoverned chaos phase is not abstract. It is documented, quantified, and, in some cases, counted in bodies.

In Myanmar, an ungoverned recommendation algorithm amplified content that contributed to the displacement of 700,000 people and the killing of thousands more. In the United States, ungoverned data collection practices enabled the Cambridge Analytica scandal — the harvesting of personal data from 87 million Facebook users without their meaningful consent, used to micro-target political advertising in the 2016 presidential election. In Ethiopia, in India, in Sri Lanka, in the Philippines, social media platforms that operated without adequate content governance amplified communal violence, political extremism, and disinformation campaigns that destabilized democratic institutions.

In the domain of privacy, the ungoverned internet produced what the scholar Shoshana Zuboff named *surveillance capitalism* — a new economic logic in which the prediction and modification of human behavior is the product, personal experience is the raw material, and the extraction of behavioral data at scale is the means of production. The data that users provided — their searches, their locations, their purchases, their relationships, their conversations, their medical concerns, their political views — was harvested, analyzed, packaged, and sold to advertisers, political campaigns, and, as the Snowden disclosures revealed in 2013, intelligence agencies. None of this required breaking the law. The law had not been written.

In the domain of labor, the internet's platforms created new categories of precarious work — the gig economy, the algorithmic management of warehouse workers, the content moderation sweatshops in which underpaid workers in the Philippines and Kenya reviewed the most violent and traumatizing content the internet produced, absorbing the psychological cost of maintaining the platforms' usability so that the platforms' users would not have to. These workers had no union, no regulatory protection specific to their conditions, and no seat in any room where the rules were being discussed.

> *"Content moderators in the Philippines and Kenya reviewed the most violent content the internet produced, absorbing the psychological cost of maintaining the platforms' usability. They had no union, no regulatory protection, and no seat in any room where the rules were being discussed."*

What the Internet Teaches AI

The internet's five-phase cycle completed in approximately fifty years — from ARPANET in 1969 to the current state of consolidated, inadequately governed platform power. This is roughly the same duration as

the electricity cycle. It is significantly faster than the printing press. And it produced a Phase 5 that arrived too late to prevent the consolidation it was supposed to constrain.

AI is following the same cycle. The parallels are not approximate. They are precise.

The innovation phase began with the deep learning breakthroughs of 2012–2020 — the moment when neural networks began achieving superhuman performance on specific tasks: image recognition, language translation, game playing. The capability existed. It was expensive, concentrated in a handful of research labs, and inaccessible to the general population.

The democratization phase began on November 30, 2022, when OpenAI released ChatGPT to the public. The system reached 100 million users in two months — the fastest adoption of a consumer technology in history, surpassing even the smartphone. Within a year, open-source models were available for anyone to download and run. Within two years, AI code generation tools were producing functional software from natural-language descriptions. The capability was distributed.

The chaos phase is now.

AI-generated disinformation is flooding information channels. AI code generation is producing software with vulnerabilities that the builders cannot detect. AI systems are being deployed in healthcare, criminal justice, military operations, and critical infrastructure without adequate testing, auditing, or oversight. The harms documented in the first book of this trilogy — biased sentencing, discriminatory healthcare, algorithmic amplification of extremism, AI-assisted targeting — are not hypothetical future risks. They are current realities.

The consolidation phase is already visible. Seven organizations can train frontier models. Two companies manufacture the required chips. Five cloud providers control the compute infrastructure.

One company controls over sixty percent of active satellites in low Earth orbit. The concentration of AI capability and infrastructure is proceeding faster than the consolidation of the internet — because AI's infrastructure requirements are more capital-intensive, more technically demanding, and more geographically concentrated than the internet's ever were.

And the governance phase? Fragments. The EU AI Act. Executive orders. National strategies. Corporate self-regulation. Voluntary commitments. No Philadelphia. No Bretton Woods. No NPT. No comprehensive, enforceable, internationally coordinated framework for the governance of artificial intelligence.

"The chaos phase is now. The consolidation is already visible. The governance is fragmentary. The internet's lesson is clear: if the room does not convene during the window, the consolidators write the rules. The window for AI is open. It is closing."

The Difference This Time

There is one crucial difference between the internet's cycle and AI's, and it cuts in both directions.

The internet's chaos phase was, in a sense, novel. Nobody had seen a global communications platform before. Nobody had experienced algorithmic amplification at scale. The ideology of internet exceptionalism could persist because there was no precedent to contradict it — no previous technology whose ungoverned chaos had produced the specific categories of harm that the internet produced.

AI does not have this excuse. We have the precedent. The internet *is* the precedent. Every failure of internet governance — the platform monopolies, the surveillance architecture, the algorithmic amplification of extremism, the erosion of privacy, the displacement of labor without safety nets, the concentration of power in five companies — is a warning, written in the lived experience of four billion people, about what happens when a transformative technology reaches consolidation without governance.

The difference this time is that the excuse is gone. We cannot claim that no one could have foreseen the consequences. We have seen them. We are living in them. The internet's ungoverned chaos is the air we breathe. If AI follows the same path — and every indicator suggests it is following the same path, faster — then the failure is not ignorance. It is choice.

The printing press took 350 years to govern because nobody had seen a printing press before. Electricity took 53 years because nobody had electrified a continent before. Nuclear energy took 23 years because nobody had split an atom before. The internet was never adequately governed, in part because nobody had built a global communications platform before.

We have built one now. We know what happened. The question is whether we will let it happen again.

Part I of this book is complete. The pattern has been established: five phases, observable across centuries, accelerating with each era, and producing consequences that are determined by the duration of the chaos phase — the period between democratization and governance. The internet is the warning case: the technology that was never adequately governed, whose chaos we are still living through, whose consolidation has locked in power structures that governance now struggles to dislodge.

Part II enters the chaos phase of artificial intelligence. Not as an abstraction. As a set of stories about specific people, in specific places, experiencing the consequences of a technology that has been democratized without governance. The new builders who create software they cannot secure. The information environment drowning in synthetic content. The workers whose judgment has been declared economically unnecessary.

The Great Convergence

The New Builders

On the developer who built a working application in a weekend, the vulnerability they could not see, and the scale of what is coming

"Maya is not a software engineer. She is not supposed to be. The tool gave her a working application. It did not give her the knowledge to evaluate whether what it gave her was safe."

The printing press was governed by popes and kings. Electricity was governed by legislatures responding to monopoly collapse. Nuclear energy was governed by heads of state who had seen what an ungoverned atom could do to a city. played out among institutions — states, corporations, militaries — and the governance response came from institutions of equivalent scale.

The chaos phase of AI is different. It is not happening exclusively in situation rooms and corporate boardrooms. It is happening in living rooms, in small offices, on laptops belonging to people who have never written a line of code and never will. The that Chapter 2 traced as Phase 2 of every technology cycle has, for AI, reached a depth and speed that has no precedent: a capability that once required a research laboratory is now available to anyone with an internet connection and a weekend. To understand why the governance window is closing, you must look at what happens when the power of a sovereign capability is handed to a physical therapist on a Sunday afternoon.

Previously, the chaos phase

The New Builders

On the developer who built a working application in a weekend, the vulnerability they could not see,and the scale of what is coming

Imagine a person — call her Maya — who runs a small physical therapy practice with three employees. She needs a system to manage patient intake, schedule appointments, and track treatment notes. The commercial software options cost between $200 and $500 per month, and none of them do exactly what she needs. She has heard about AI tools that can build applications from plain-language descriptions. On a Saturday morning, she opens one of these tools and types: "Build me a patient management system for a small physical therapy practice."

By Sunday evening, she has a working application. It has a patient intake form, a scheduling calendar, a treatment notes interface, and a dashboard. It looks professional. It functions. Her employees can use it on Monday. She is delighted. She has saved thousands of dollars in software costs and built something tailored to her practice's specific needs.

The application stores patient health records — protected health information under federal privacy laws — HIPAA in the United States, PIPEDA in Canada, GDPR in Europe — regulations that impose strict requirements on how such data is stored, accessed, and secured. The AI tool that generated the code did not implement encryption at rest for the database. It did not implement proper session management. It did not implement role-based access controls. It did not include audit logging. It stored the database connection credentials in a configuration file accessible from the web. Maya does not know what any of these things are. The AI tool did not tell her they were missing, because the AI tool was not designed to audit its own output for security or regulatory compliance. It was designed to produce functional code from natural-language descriptions. Functional and secure are not the same thing.

Maya is not a software engineer. She is not supposed to be. She is a physical therapist who needed a scheduling system. The tool gave her one. The tool did not give her the knowledge to evaluate whether what it gave her was safe. And no governance framework requires her to have that knowledge, requires the tool to provide that evaluation, or requires anyone to stand between the tool's output and the patient data it will process.

In 2025, security researchers identified a critical vulnerability — CVE-2025-48757 — in applications built using Lovable, one of the most popular AI application builders. The vulnerability exposed the data of applications built on the platform to unauthorized access. The security assessment found that approximately one in ten applications built with the tool were affected. One in ten.

The vulnerability was not in the users' code. It was in the patterns the AI generated — the default configurations, the architectural choices, the security assumptions embedded in the code the tool produced. The users who built applications with Lovable did not introduce the vulnerability. They inherited it. They inherited it from a tool they trusted, in code they could not read, implementing security practices they could not evaluate.

The reader who has followed this trilogy will recognize the architecture. In *The Last Keeper*, Chapter 3, the Therac-25's software was reused from the Therac-20. The software contained a flaw that was harmless on the old machine, where hardware interlocks caught it, and lethal on the new machine, where the interlocks had been removed. In *The Last Keeper*, Chapter 6, the Ariane 5 rocket reused software from the Ariane 4. The code worked exactly as written — for the wrong rocket. The architecture recurs: code produced in one context, deployed in another, carrying assumptions that are invisible to the person who deploys it.

The difference in scale is the difference that demands governance. The Therac-25 was deployed in a handful of clinics. The Ariane 5 was a single rocket. Lovable and tools like it are used by tens of thousands of builders, producing applications that serve millions of users, handling data that includes medical records, financial information, personal identification, and communications. The inherited flaw scales with the tool's adoption. The Therac-25's flaw harmed six patients. A flaw in an AI code generation pattern can potentially affect every application built with that pattern.

The Paradox

In 2025, the nonprofit research organization METR — Model Evaluation and Threat Research — published a study that measured the actual productivity impact of AI coding tools on experienced software developers working on real-world projects. The finding was striking: developers using AI assistance were 19% slower than developers working without it. At the same time, the developers using AI assistance reported feeling 20% faster.

The gap between perceived and actual performance is not a curiosity. It is a safety problem. A developer who believes they are working faster and more effectively will rely more heavily on the tool. They will review the tool's output less carefully. They will trust the generated code more readily. And the code they trust will contain the assumptions, the defaults, and the blind spots of the model that generated it — assumptions and blind spots that the developer, feeling confident and productive, is less likely to question.

> *"Maya is not a software engineer. She is not supposed to be. The tool gave her a working application. It did not give her the knowledge to evaluate whether what it gave her was safe. Nobody stands between the tool's output and the patient data."*

The Vulnerability

Maya is fictional. The vulnerability is not.

The Last Keeper, Chapter 4 — the pilots of Air France 447 — translated from the cockpit to the code editor. The more reliable the autopilot, the less the pilot flew manually, the worse the pilot performed when the autopilot disconnected. The more capable the AI coding tool, the less the developer exercises independent judgment about the code, the worse the developer performs at catching the tool's errors. The tool is good enough that the developer stops checking. The developer stops checking precisely because the tool is good enough. And then the tool produces a vulnerability, and nobody is watching.

> *"Developers using AI assistance were 19% slower but felt 20% faster. The gap between perceived and actual performance is not a curiosity. It is the automation paradox, translated from the cockpit to the code editor."*

The Mandate

In early 2025, Tobi Lütke, the CEO of Shopify, sent an internal memo to his company that was subsequently made public. The memo stated that the use of AI tools was now a baseline expectation for every employee at the company. Not an option. Not an experiment. A baseline. Employees would be evaluated on their ability to leverage AI effectively. Teams requesting additional headcount would first need to demonstrate that they had explored what AI could do in place of a new hire.

Shopify is one of the largest e-commerce platforms in the world, serving over 2 million merchants globally. The memo was not a suggestion from a startup founder. It was a policy directive from the CEO of a company whose infrastructure underpins a significant fraction of global online commerce. And it was representative of a broader shift: across the technology industry, AI adoption moved in 2025 from exploration to mandate. Companies did not ask employees whether they wanted to use AI. They told them to.

The implications extend beyond Shopify's employees. The merchants on Shopify's platform — the small businesses, the independent sellers, the entrepreneurs — would increasingly interact with AI-generated tools, AI-generated interfaces, and AI-generated business logic built by a workforce that had been instructed to rely on AI as a default. The applications that Maya the physical therapist builds with AI tools are one layer of the problem. The platforms on which those applications run, themselves built with increasing AI assistance, are another. The assumption of human review at every layer of the software stack is no longer valid.

The Lego and the Lumber

There is a way to understand the landscape that this chapter describes, and it is the distinction between two approaches to building software that will define the governance challenge for the next decade.

The first approach is model-driven development — building software on platforms that enforce structure, constraints, and guardrails by design. The bricks snap together in predefined ways. You can build complex structures, but the bricks constrain what you can build to configurations that the platform has validated. The brick cannot be assembled in a way that produces an insecure database connection, because the platform does not offer that assembly. The safety is in the brick, not in the builder.

The second approach is code-generated development — building software by generating raw code from AI tools, with no platform enforcing structure or constraints. You can build anything. The lumber does not constrain you. But the lumber does not protect you, either. If you do not know how to frame a load-bearing wall, the house falls down. If you do not know how to implement encryption at rest, the patient data is exposed. The capability is in the tool. The safety is in the builder's knowledge. And the builder, increasingly, does not have that knowledge — because the tool has made the knowledge feel unnecessary.

> *"Model-driven development is Lego: the safety is in the brick. Code-generated development is raw lumber: the safety is in the builder's knowledge. The tool has made the knowledge feel unnecessary."*

The governance challenge is this: AI code generation is expanding the population of builders at an rate. People who could not build software a year ago can now produce functional applications in hours. This is genuinely democratizing. It is genuinely valuable. And it is genuinely dangerous, because the new builders are building with lumber — raw, unstructured, unconstrained code — without the expertise to know whether what they have built is sound.

The printing press democratized the production of text. The chaos phase was fueled by the fact that anyone could now publish anything — including inflammatory pamphlets that incited religious warfare. AI code generation is democratizing the production of software. The chaos phase will be fueled by the fact that anyone can now build anything — including applications that handle sensitive data without the security practices that a professional developer would consider non-negotiable.

The Scale of What Is Coming

Maya's physical therapy practice is one case. The scale of the phenomenon is vastly larger.

In the United States alone, there are approximately 33 million small businesses. Globally, the number exceeds 400 million. The vast majority of these businesses do not employ software developers. They have historically relied on commercial software — purchased applications built by professional development teams, tested for security and compliance, and maintained through regular updates. The commercial software was expensive and generic. But it was, in most cases, professionally built. remarkable rate. Like building with Lego bricks.

Like building with raw lumber.

AI code generation changes this equation. Within the next five years, an estimated twenty million or more small businesses, nonprofits, and individuals will use AI tools to build custom software for their specific needs. They will build patient management systems. They will build customer databases. They will build payment processing interfaces. They will build tools that handle financial records, medical histories, educational data, and personal communications. They will build these tools because AI has made building possible for people who could not build before. And they will deploy these tools without security audits, without compliance reviews, without penetration testing, and without the professional oversight that the commercial software they are replacing would have received.

The comparison to the printing press is precise. Before Gutenberg, the production of text was controlled by trained scribes operating within institutional frameworks — monasteries, universities, royal courts — that provided a degree of quality control and institutional accountability. After Gutenberg, anyone with access to a press could produce text. The quality, accuracy, and responsibility of that text became the individual printer's concern. Some printers were careful. Many were not. The governance frameworks that eventually addressed this — copyright law, libel law, press standards — took centuries to develop.

Before AI code generation, the production of software was controlled by trained developers operating within professional frameworks — engineering teams, code review processes, security audits, compliance certifications — that provided a degree of quality control and institutional accountability. After AI code generation, anyone with access to the tool can produce software. The security, compliance, and reliability of that software becomes the individual builder's concern. Some builders will be careful. Many will not know what careful means.

"Before AI code generation, software was produced by trained developers within professional frameworks. After AI code generation, anyone can produce software. Some builders will be careful. Many will not know what careful means."

The Missing Keeper

In the old model, there were keepers in the software development process. Senior engineers who reviewed code before it was deployed. Security teams that audited applications before they went live. Compliance officers who verified that systems handling regulated data met the applicable requirements. QA testers who looked for the failure modes the developers did not anticipate.

These keepers were expensive. They were slow. They were, from the perspective of a small business trying to ship a product, an obstacle. But they were the mechanism by which software moved from "it works" to "it works and it is safe." The distance between those two statements is the distance between Maya's application and an application that a professional team would deploy to handle patient health records.

AI code generation removes these keepers — not by replacing them with something better, but by making them seem unnecessary. The tool produces working code. Working code feels like finished code. The builder who has never worked with a security team does not know that a security review is missing. The builder who has never seen a compliance audit does not know that a compliance audit is needed. The absence of the keeper is invisible to the person who has never seen the keeper work.

This is a governance problem, not a technology problem. The technology is doing what it was designed to do: producing functional software from natural-language descriptions. The governance question is: who ensures that the software is safe? Who stands between the tool's output and the data it will process? Who catches the vulnerability that the builder cannot see and the tool does not flag?

In the current landscape, the answer is: nobody. No regulatory framework requires AI-generated software to be audited before deployment. No licensing requirement ensures that the person building software with AI tools understands the security implications of what they are building. No certification standard applies to the AI tool itself, requiring it to produce code that meets minimum security or compliance benchmarks. No governance framework exists for this scenario. And twenty million builders are about to build.

"No regulatory framework requires AI-generated software to be audited before deployment. No certification standard applies to the AI tool itself. No governance framework exists for this scenario. And twenty million builders are about to build."

What Governance Would Look Like

This chapter is not an argument against AI code generation. The democratization of building is genuine and valuable. Maya should be able to build a patient management system that serves her practice. The question is not whether she should build. The question is what the governance framework should require so that what she builds does not expose her patients to harm.

The analogy is to building codes — the governance framework that emerged from the chaos phase of physical construction. Anyone can design a house. Not anyone can build one that meets code. Building codes do not prevent people from building. They ensure that what is built meets minimum standards for safety, structural integrity, and habitability. They apply to the builder, not to the homeowner's intentions. A homeowner who builds a deck that collapses is responsible for building without a permit. A jurisdiction that does not require a permit is responsible for the absence of the requirement.

The software equivalent of building codes does not exist for AI-generated applications. It could. It would require AI code generation tools to include automated security auditing as a default — not as an optional add-on, but as an integral part of the generation process. It would require applications that handle regulated data to pass automated compliance checks before deployment. It would require the platforms that host these applications to verify minimum security standards, just as building departments verify

minimum construction standards. It would require the AI tool providers to be accountable for the security patterns their tools produce, just as lumber mills are accountable for the grade of the lumber they sell.

None of this is technically impossible. Automated security scanning tools exist. Compliance verification systems exist. The technology to make AI code generation safer is available. What is absent is the governance framework that requires it to be used. The building codes have not been written. The room has not convened. And every day that the room does not convene, more applications are built, more data is exposed, and more patients, customers, and users bear the risk of a regulatory void they did not create and cannot see.

The new builders are building software without governance. The next chapter describes a parallel chaos: the new content creators are flooding the information environment with synthetic material that is indistinguishable from reality. AI has democratized not just the production of code but the production of deception — and the cost of determining what is true is rising faster than any institution can bear.

The Great Convergence

The Information Flood

On the democratization of deception, and the epistemic crisis that arrives when fabrication is free and detection is expensive

"When any image might be fabricated, trust in all images erodes. The damage is not limited to the specific fabrication. The damage is to the shared capacity of a society to agree on what is real."

The Information Flood

On the democratization of deception, and the epistemic crisis that arrives when determining what is true costs more than anyone can afford

In 1835, the *New York Sun* published a series of articles claiming that Sir John Herschel, one of the most prominent astronomers of the era, had observed life on the Moon through a powerful new telescope. The articles described bat-winged humanoids, blue unicorns, and a civilization of temples and obelisks. The series, which became known as the Great Moon Hoax, was entirely fabricated. It was written by a reporter named Richard Adams Locke, and its purpose was to sell newspapers. It worked. The *Sun* became the bestselling newspaper in the world.

The Great Moon Hoax required a writer, an editor, a printing press, and a distribution network. It required weeks of planning and composition. It required the reputation of a real astronomer to lend credibility. And it was debunked within weeks by rival publications that had the time, the expertise, and the institutional motivation to investigate the claims.

What happens when the cost of producing a convincing fabrication drops to zero and the infrastructure required to debunk it does not change. The Great Moon Hoax required a newsroom. A deepfake video requires a laptop and a text prompt. The asymmetry between the cost of producing deception and the cost of identifying it is the structural dynamic that defines the information environment of the AI era — and it is a dynamic that no existing governance framework addresses.

> *"The Great Moon Hoax required a newsroom. A deepfake video requires a laptop and a text prompt. The asymmetry between the cost of producing deception and the cost of identifying it is the crisis."*

The Economics of Deception

Every form of deception has an economics — a cost of production and a cost of detection. The history of disinformation is the history of that ratio shifting.

Before the printing press, producing a convincing forgery required a skilled scribe, expensive materials, and access to the formats and seals of whatever authority the forgery claimed to represent. The cost of production was high. The cost of detection was also high, but the two were roughly in balance — a forgery that took weeks to produce could be investigated over a similar timescale.

The printing press shifted the ratio. Mass-produced pamphlets could spread fabricated claims faster than any authority could respond. The inflammatory pamphlets that fueled the Wars of Religion were, in their era, the equivalent of viral misinformation — produced cheaply, distributed widely, and consumed by populations that had no institutional mechanism for evaluating their accuracy. The chaos phase of the printing press was, in significant part, a crisis of information governance.

Photography and film shifted the ratio again. A photograph was assumed to be a reliable representation of reality. Manipulating a photograph required darkroom skill and was detectable by experts. A film required actors, sets, cameras, editing equipment, and distribution infrastructure. The cost of producing convincing visual deception remained high enough that detection could keep pace.

Digital editing tools shifted the ratio further. Photoshop made image manipulation accessible to anyone with a computer. Video editing tools made it possible to alter footage. But professional forensic analysis could still detect the alterations — inconsistencies in lighting, compression artifacts, metadata that revealed the editing history. The cost of production dropped. The cost of detection remained manageable.

Generative AI has broken the ratio.

The Break

In 2024 and 2025, AI systems capable of generating photorealistic images, realistic video, and natural-sounding audio became widely available — many of them free, open-source, and runnable on a consumer laptop. The cost of producing a convincing fabrication — a deepfake video of a political figure saying something they never said, a synthetic audio recording of a CEO announcing a merger that does not exist, a fabricated photograph of an event that never happened — dropped from thousands of dollars and specialized expertise to zero dollars and a text prompt.

The cost of detecting these fabrications did not drop. It increased. Deepfake detection requires specialized AI models trained specifically to identify synthetic content, forensic tools that analyze pixel-level artifacts, and human expertise to interpret the results. The detection tools lag behind the generation tools — the generators improve faster than the detectors can adapt, because generating convincing content requires solving one problem (make it look real) while detecting it requires solving many (identify every possible artifact of every possible generation method).

The asymmetry is now structural. One person with a laptop can produce a hundred pieces of synthetic content in a day. A newsroom, a fact-checking organization, or a platform's trust-and-safety team must spend hours or days verifying each one. The production scales effortlessly. The detection does not. And every piece of synthetic content that enters the information environment — whether or not it is detected — raises the ambient level of uncertainty for everything else. When any image might be fabricated, trust in all images erodes. When any audio might be synthetic, trust in all audio erodes. The damage is not limited to the specific fabrication. The damage is to the epistemic infrastructure itself — the shared capacity of a society to agree on what is real.

The Elections

The 2024 US presidential election cycle was the first major democratic exercise conducted in an information environment saturated with generative AI. Deepfake audio clips of candidates circulated on social media. AI-generated images depicting events that never happened were shared as evidence. Synthetic robocalls mimicking the voice of a sitting president were used to discourage voters from going to the polls in a primary election.

The robocall incident illustrates the asymmetry at human scale. In January 2024, two days before the New Hampshire primary, between 5,000 and 25,000 registered Democrats received a phone call in which a voice indistinguishable from President Biden's told them not to vote. "Save your vote for November," the synthetic voice said. "Voting this Tuesday only plays into the hands of the Republicans." The call was traced to a political operative who had paid a mere $150 to generate the deepfake audio using a commercially available AI voice-cloning tool. One hundred and fifty dollars to suppress the votes of thousands of citizens in a presidential primary. The cost of producing the deception: trivial. The cost of identifying, investigating, and prosecuting it: months of work by the state attorney general's office, the FCC, and federal investigators.

The United States was not unique. In elections across the world — in India, Indonesia, the United Kingdom, South Korea, and dozens of other democracies that held elections in 2024 and 2025 — AI-generated content was documented as a tool of political manipulation. In some cases, the content was produced by domestic political actors. In others, by foreign governments seeking to influence outcomes. In many cases, the origin could not be determined — because the tools that produce the content do not leave fingerprints the way a newsroom or a government propaganda bureau does.

The governance response was fragmented and reactive. Some jurisdictions introduced requirements for AI-generated content to be labeled. The requirements were largely unenforceable — because labeling is voluntary for content shared on open platforms, because the tools that generate the content do not embed mandatory watermarks, and because the platforms that distribute the content have neither the technical capacity nor the economic incentive to verify the provenance of every piece of media that passes through their systems.

This is the content moderation failure, amplified by an order of magnitude. In *The Last Keeper*, Chapter 8, Facebook's fewer than two dozen Burmese-speaking moderators were overwhelmed by a platform serving twenty million users. The AI content generation tools now available to anyone can produce synthetic content at a rate that would overwhelm every fact-checking organization on earth working simultaneously. The Speed-Judgment Tradeoff from that chapter — the system operating faster than any human review can match — has crossed into a domain where the content itself is machine-generated. The machine produces the content. The machine distributes the content. And the

"When any image might be fabricated, trust in all images erodes. The damage
is not limited to the specific fabrication. The damage is to the shared capacity
of a

human who is supposed to evaluate the content is outpaced at both ends.

The Epistemic Crisis

The term "epistemic crisis" describes a situation in which the shared mechanisms by which a society determines what is true have degraded to the point that consensus on basic facts becomes impossible. — philosophers of science and political theorists have discussed epistemic fragility for decades. What is new is the mechanism by which the crisis is arriving.

Previous epistemic crises were produced by the intentional actions of identifiable actors: state propaganda, institutional deception, organized disinformation campaigns. The actors could, at least in principle, be identified, confronted, and held accountable. The crisis could be attributed. The counter could be organized.

The AI-driven epistemic crisis is different in kind. It is produced not by a coordinated campaign but by the structural economics of information production. When the cost of producing content approaches zero, the information environment is flooded regardless of anyone's intent. A teenager generating deepfakes for social media followers. A content farm producing thousands of AI-written articles to capture advertising revenue. A political operative generating a few hundred synthetic images to shift a narrative. An AI chatbot confidently providing medical information that is factually wrong. None of these actors is running a coordinated disinformation campaign. All of them, collectively, are degrading the information environment in ways that no single actor can be held responsible for.

The result is not that people believe lies. The result is worse: people stop believing anything. When the information environment is sufficiently polluted, the rational response is not to trust more carefully but to trust less universally. Citizens disengage from news. Voters distrust electoral outcomes. Patients question medical guidance. Juries doubt video evidence. The foundation of shared factual reality on which democratic governance, public health, and the administration of justice depend begins to dissolve — not because the facts have changed but because the mechanism for agreeing on the facts has been overwhelmed.

> *"The result is not that people believe lies. The result is worse: people stop believing anything. The mechanism for agreeing on facts has been overwhelmed. And when a society cannot agree on what is real, it cannot govern itself."*

The Liar's Dividend

There is a second-order effect that may be more damaging than the fabrications themselves. Researchers have named it the liar's dividend: the benefit that accrues to liars when the existence of deepfakes allows them to dismiss authentic evidence as fabricated. Not a new concept

Before deepfakes, a video of a public figure committing an act or making a statement was difficult to deny. The video existed. It could be authenticated. It was evidence. In the deepfake era, anyone confronted with authentic video evidence can claim it was AI-generated. The claim does not need to be proven. It only needs to be plausible — and in an environment where deepfakes are known to exist, it is always plausible.

The liar's dividend inverts the burden of proof. In a pre-deepfake information environment, the burden was on the person claiming a video was fake to demonstrate the fabrication. In a post-deepfake environment, the burden shifts to the person presenting the video to prove it is authentic — a proof that may be technically possible for forensic experts but is practically impossible for the average citizen, the journalist on deadline, or the jury evaluating evidence.

The liar's dividend benefits the powerful at the expense of the accountable. A politician caught on video making a damaging statement can claim the video is a deepfake. A corporation whose internal practices are documented on video can challenge the footage's authenticity. A government whose security forces are filmed committing atrocities can dismiss the evidence as AI-generated propaganda. In each case, the existence of deepfake technology provides plausible deniability that did not previously exist — not because the specific evidence is fabricated, but because the environment in which it is presented no longer assumes that video evidence is real.

for the information domain. The keeper — the journalist, the fact-checker, the forensic analyst, the investigative body — can still do the work. The work still produces accurate results. But the environment in which the results are presented has been degraded to the point where the results are dismissible. The keeper is not removed from the loop. The loop itself has been made unreliable.

> *"The liar's dividend inverts the burden of proof. Anyone confronted with authentic evidence can claim it was AI-generated. The claim does not need to be proven. It only needs to be plausible. In the deepfake era, it is always plausible."*

The Governance Gap

The governance challenge here is not content moderation — the approach the internet's platforms have pursued, with limited success, for two decades. Content moderation addresses the symptom: harmful content that has already been produced and distributed. The governance challenge of generative AI addresses the infrastructure: the tools that produce the content, the models that generate it, and the absence of any framework that governs their use.

Several governance approaches have been proposed. Content provenance systems — technologies that embed verifiable metadata in authentic media at the point of capture, allowing downstream consumers to verify that an image or video was captured by a real camera and has not been altered — are The keeper's problem restated

technically promising. The Coalition for Content Provenance and Authenticity — known as C2PA — has developed an open standard for this purpose. But adoption requires cooperation across the entire media chain: camera manufacturers, software developers, social media platforms, and news organizations. No governance framework mandates adoption. The standard is voluntary. And voluntary standards in an environment where the economic incentive favors speed over verification have a poor track record.

Mandatory watermarking of AI-generated content has been proposed and, in some jurisdictions, enacted. The technical challenge is that watermarks can be stripped or degraded by reprocessing the content — screenshotting an image, re-encoding a video, or passing text through a paraphrasing tool. Watermarking addresses content produced by compliant tools. It does not address content produced by tools that are open-source, self-hosted, or operated by actors who choose not to comply. The governance reaches the willing. The willing are not the problem.

The deeper governance challenge is the one this book has traced since the printing press: the technology has democratized a capability, the capability produces consequences that existing governance cannot contain, and the consolidation of the tools in a small number of companies creates a structural tension between governing the tool (which the companies resist) and governing the content (which is impossible at scale). No governance framework addresses AI-generated information any more than it convened for the internet. And the chaos is accumulating at a pace that makes the internet's thirty-year governance gap look leisurely.

The Parallel

The previous chapter described the chaos in software: millions of people building applications without the knowledge to evaluate their safety. This chapter describes the chaos in information: billions of people navigating an environment in which the tools to create convincing content have been given to everyone and the tools to verify that content have been given to almost no one.

The two are not separate problems. They are the same problem — the Phase 3 chaos of AI's five-phase cycle, expressing itself simultaneously across multiple domains. In software, the chaos takes the form of ungoverned code. In information, the chaos takes the form of ungoverned content. In both cases, the mechanism is identical: a capability that was previously controlled by professionals within institutional frameworks has been democratized without the governance frameworks that made the professional version safe.

The next chapter completes the triptych. Software. Information. And now: labor. The third domain in which AI's democratization is producing chaos — not because the technology is failing, but because the technology is succeeding, and the success is displacing the people whose judgment the previous books argued was irreplaceable.

The Great Convergence

The Power Shift

On the displacement of judgment, and the governance vacuum for a labor transformation without precedent

"The question is not whether AI will displace human labor. It is whether the displacement will be governed."

The Power Shift

On the displacement of judgment, and the governance vacuum for a labor transition that is moving faster than any institution can adapt

In 2024, Klarna, the Swedish financial technology company, announced that its AI systems were performing work that had previously required approximately 700 full-time employees. The announcement was made not with regret but with pride — as evidence of the company's technological sophistication and operational efficiency. The AI handled customer service inquiries, processed claims, resolved disputes, and performed the routine analytical work that had been the daily labor of seven hundred human beings.

Seven hundred is a number. Let us pause on what seven hundred means.

Seven hundred people woke up each morning and went to work. They had skills they had developed over years. They had expertise in Klarna's products, its policies, its customer base. They understood the nuances that distinguish a legitimate complaint from a fraudulent one, a confused customer from an angry one, a billing error from a billing dispute. They exercised judgment — the specific human capability that this trilogy has argued, across two books and twenty-six chapters, is irreplaceable in consequential domains. And they were replaced. Not by better-trained replacements. Not by a reorganization that shifted their work elsewhere. By a system that their employer determined could do their work at a fraction of the cost, at greater speed, with acceptable quality.

The announcement did not name the seven hundred. It did not describe their transition — whether they were reassigned, retrained, or simply let go. The number was a data point in a corporate efficiency narrative. The people behind the number were invisible.

> *"Seven hundred people exercised judgment every day. They understood the nuances that distinguish a confused customer from an angry one, a billing error from a dispute. They were replaced by a system that their employer determined could do the work faster and cheaper. The people behind the number were invisible."*

The Difference This Time

Every major technology transition in the history of industrialization has displaced labor. The mechanization of agriculture eliminated millions of farm jobs. The factory system displaced artisans and cottage workers. The assembly line concentrated production and eliminated skilled trades. Containerization destroyed the longshoreman's craft. Computerization automated clerical work. In each case, the transition was painful, often brutal, and eventually — over decades — the labor market adapted. New industries absorbed the displaced workers, or their children, or their grandchildren. The adaptation was not kind. It was not fast. But it happened.

The standard reassurance about AI displacement follows this script: technology has always displaced jobs and always created new ones. The printing press destroyed the scribe's profession and created the publishing industry. The automobile destroyed the horse-drawn carriage industry and created the automotive sector. Computerization destroyed typing pools and created the software industry. AI will destroy some jobs and create others. The transition will be painful but temporary.

This reassurance is based on a pattern that held not hold for this one. The reason is a difference that the standard reassurance does not address: every previous automation wave displaced hands. This one displaces judgment.

The mechanization of agriculture displaced physical labor — the act of planting, harvesting, and processing crops by hand. The assembly line displaced manual skill — the act of shaping, assembling, and finishing products by hand. Computerization displaced clerical work — the act of recording, calculating, and filing information by hand. In each case, the work that was automated was execution: the performance of a defined task according to defined procedures. The judgment about what to do, when to do it, and whether the result was correct remained with the human.

AI automates the judgment itself. A customer service AI does not merely transcribe the customer's complaint. It evaluates the complaint, determines the appropriate resolution, and executes it. A legal AI does not merely search for relevant case law. It analyzes the case, assesses the arguments, and drafts the brief. A medical AI does not merely display the scan. It reads the scan, identifies the anomaly, and recommends the diagnosis. The function being automated is not the hand but the mind — not the execution of a decision but the making of it.

> *" — the execution of a defined task. AI displaces the mind — the making of the decision. The function being automated is not execution but judgment."*

The Keepers Displaced

The connection to the first book of this trilogy is direct and troubling.

The Last Keeper spent twelve chapters arguing that human judgment in consequential systems is irreplaceable — that removing the human from the loop produces catastrophe at the margin, that the automation paradox degrades the human's ability to catch the system's errors, and that the keeper must remain. The argument was about safety. It was about the patient on the table, the passenger in the aircraft, the defendant in the courtroom.

The labor market is now removing the keepers for a different reason: not safety but economics. The seven hundred Klarna employees were not removed because the AI was safer than they were. They were removed because the AI was cheaper. The economic logic that drives AI labor displacement is indifferent to the safety argument. A company that replaces human judgment with AI judgment is not making a safety prior automation waves displaced the hand for prior automation waves. It may

decision. It is making a cost decision. And the cost calculation does not include the cost documented in *The Last Keeper* — the cost that arrives at the margin, when the system encounters conditions it was not trained for, and the human who would have caught the error is no longer employed.

The safety argument says: keep the human in the loop. The economic argument says: remove the human from the payroll. The two arguments are not debating each other in a conference room. They are operating simultaneously, in the same organizations, on the same systems, with the economic argument winning consistently because the economic argument produces measurable quarterly savings and the safety argument produces unmeasurable future risk. The savings are certain and immediate. The risk is probabilistic and deferred. In every organization governed by quarterly metrics, the certain and immediate wins.

> *"The safety argument says: keep the human in the loop. The economic argument says: remove the human from the payroll. The savings are certain and immediate. The risk is probabilistic and deferred. In every organization governed by quarterly metrics, the certain wins."*

The Speed

The displacement is not only different in kind from previous waves. It is different in speed.

The mechanization of agriculture in the United States unfolded over roughly a century — from the mid-nineteenth century to the mid-twentieth. The transition from horse-drawn farming to mechanized farming took three generations. Workers displaced from farms migrated to cities, found factory work, and their children entered the service economy. The adaptation was measured in decades and generations.

The computerization of clerical work unfolded over roughly forty years — from the introduction of the personal computer in the early 1980s to the present. Typing pools disappeared. Filing clerks were replaced by databases. Bookkeepers were replaced by spreadsheets. The displaced workers retrained, found adjacent roles, or aged out of the workforce. The adaptation was measured in decades.

AI labor displacement is measured in months. Klarna's announcement covered a transition that occurred within a single year. The capabilities of AI systems are improving not on annual release cycles but on quarterly or monthly ones. A task that AI cannot perform in January may be automated by June. A role that is secure in Q1 may be eliminated in Q3. The pace of displacement exceeds the pace of any institutional response — retraining programs, educational pipelines, social safety nets, labor market adjustments — by an order of magnitude.

No governance framework exists for labor displacement at this speed. Previous transitions, painful as they were, unfolded slowly enough that institutions could adapt. Labor unions formed. Governments enacted minimum wage laws, unemployment insurance, job retraining programs. Social safety nets were built — imperfectly, inadequately, but built — over the decades that the transitions required. AI The convergence that the title names.

displacement does not provide decades. It provides months. And the governance response, which moves at the speed of legislation, regulation, and institutional adaptation, cannot match the speed of the technology that is producing the displacement.

The Paradox of Democratization

There is a paradox at the center of this chapter, and it connects directly to the argument of *Who Holds the Jar?*.

AI is simultaneously democratizing capability and concentrating economic value. Maya, the physical therapist from Chapter 4, can now build software she could not build before. A student can now produce analysis that previously required a team of analysts. A small business can now generate marketing materials, legal documents, financial models, and customer communications that previously required specialized professionals. The capability has been distributed. Anyone can do what previously required expertise.

But the economic value of the capability flows not to the new users but to the companies that provide the tools. OpenAI, Anthropic, Google, Microsoft, and the other companies that build and host AI systems capture the subscription fees, the API revenue, the enterprise contracts, and the data generated by billions of interactions. The professionals whose expertise has been democratized — the analysts, the designers, the writers, the rs, the customer service agents — have seen their economic value diminish, because the scarcity that justified their compensation has been eliminated. The tool that gives everyone the capability takes away the premium that the capability once commanded.

followed by reconcentration from *Who Holds the Jar?* operating at the level of human labor. The capability distributes. The economic value recaptures. The users feel empowered. The infrastructure owners accumulate the wealth. And the professionals who previously held the capability as a source of livelihood find that their expertise has been commoditized by the same tool that made it accessible to everyone else.

> *"AI simultaneously democratizes capability and concentrates economic value.*
> *The capability distributes. The wealth recaptures. The users feel empowered.*
> *The professionals find their expertise commoditized."*

The Professions

Klarna's customer service agents were the most visible displacement. They will not be the most consequential.

In April 2025, Shopify CEO Tobi Lütke sent an internal memo that leaked within hours. The message was blunt: before any team could request additional headcount, they would first need to demonstrate that AI could not do the job. The memo reframed hiring itself as a failure of imagination — the default was now

program

that the work should be done by the machine, and a human was the exception that required justification. Lütke was not describing a future policy. He was describing the present. Within weeks, other technology companies adopted similar frameworks, and the phrase "prove the machine can't" entered the vocabulary of workforce planning across the industry.

In law, AI systems are already capable of reviewing contracts, conducting legal research, drafting briefs, and summarizing depositions — tasks that constitute the daily work of junior associates and paralegals at law firms worldwide. The profession that trains its young practitioners by assigning them precisely this work is discovering that the training pipeline and the automation target are the same thing. The junior lawyer who would have spent three years reviewing contracts — and in doing so, learning to think like a lawyer — will not have those three years. The work will be done by the machine. And the senior lawyer who was produced by that training pipeline will not be produced by the next one.

In medicine, AI diagnostic systems are approaching or exceeding the accuracy of human specialists in specific domains — radiology, dermatology, pathology. The radiologist who spent a decade developing the pattern recognition skills that allow them to read a scan and see the anomaly that others miss is being told that the machine can do it faster and, in controlled studies, more accurately. The studies may be correct. The question — examined across *The Last Keeper*, — is what happens when the machine is wrong and the radiologist who would have caught the error has not been trained, because the machine made the training seem unnecessary.

In finance, AI systems are conducting analysis, generating reports, making trading decisions, and assessing risk at speeds and volumes that human analysts cannot match. In journalism, AI is producing news summaries, sports recaps, earnings reports, and weather forecasts. In education, AI tutoring systems are providing personalized instruction. In architecture, AI is generating building designs. In accounting, AI is preparing tax returns and conducting audits.

The pattern is consistent across every profession: the work that is most amenable to AI automation is the work that constitutes the training pipeline for the profession's future practitioners. The junior work

— the document review, the preliminary analysis, the first-draft writing, the routine diagnostics — is both the most automatable and the most formative. Automating the junior work accelerates the present and hollows out the future. The firm is more efficient today. The firm has no pathway to develop the senior expertise it will need in ten years.

"The work most amenable to AI automation is the work that constitutes the training pipeline for the profession's future practitioners. Automating the junior work accelerates the present and hollows out the future."

The Governance Vacuum

When the mechanization of agriculture displaced millions of farm workers, the governance response — slow, partial, and often inadequate — included agricultural subsidies, rural development programs, the expansion of public education to prepare the next generation for industrial work, and eventually the social safety net programs of the New Deal. These responses took decades to develop and never fully addressed the displacement. But they existed. They were written. The room convened, however belatedly.

When computerization displaced clerical workers, the governance response included workforce retraining programs, community college expansion, and the development of technology education in public schools. Again, inadequate. Again, slow. But present.

For AI-driven labor displacement, as of this writing, the governance response is: almost nothing.

No major economy has enacted legislation specifically addressing AI labor displacement. No comprehensive retraining framework has been established at a scale proportional to the displacement that is already occurring. No social safety net adaptation has been designed for a labor market in which the pace of job elimination exceeds the pace of job creation for the first time in the history of industrialization. The proposals that exist — universal basic income, job guarantee programs, AI taxation to fund transition support — remain proposals. They are discussed in academic papers, debated at conferences, and advocated by think tanks. They are not law. They are not policy. They are not governance.

The governance response is: almost nothing. And the seven hundred are already gone.

The Chaos Phase Restated

the chaos phase of artificial intelligence across three domains.

In Chapter 4, the chaos in software: millions of people building applications without the expertise to evaluate their safety, using tools that produce functional code without producing secure code, in a landscape where no governance framework requires the tools or the builders to meet minimum standards.

In Chapter 5, the chaos in information: the cost of producing convincing fabrication has dropped to zero while the cost of detecting it has not, producing an epistemic crisis in which the shared capacity of societies to agree on basic facts is being overwhelmed, and the liar's dividend allows the powerful to dismiss authentic evidence as fabricated.

In this chapter, the chaos in labor: AI is displacing not hands but judgment, at a speed that exceeds the capacity of any institutional response, hollowing out the training pipelines that produce the next generation of expertise, and concentrating the economic value of the democratized capability in the companies that provide the tools.

Software. Information. Labor. Three domains. One pattern: a capability has been democratized without the governance frameworks that would make the democratization safe for the people it affects. The tools are available. The rules are not. The chaos is the gap between the two. Part II traced

Part III asks the question that follows: in the absence of governance, who is consolidating control? Who is gaining the power that the chaos phase produces? And what are they doing with it?

> *"Software. Information. Labor. Three domains. One pattern: a capability has been democratized without the governance frameworks that would make the democratization safe. The tools are available. The rules are not. The chaos is the gap."*

The Great Convergence

The New Gatekeepers

On the consolidation of infrastructure, the concentration of power, and the entities that now control the gates

"Seven organizations can train a frontier model. Two companies manufacture the chips. Five providers run the cloud. One constellation controls the orbit. This is not a market. It is a chokepoint."

The New Gatekeepers

On the night drone submarines lost connectivity in the Black Sea because one man decided they should, the seven companies that can train frontier AI models, and the discovery that the consolidation of AI infrastructure is already more concentrated than any previous technology in history

"A single private individual's decisions about a commercial satellite network are now a primary variable in a major conventional war. No governance framework covers this situation. No governance framework was designed to."

The New Gatekeepers

On the consolidation of infrastructure, the concentration of power, and the person who holds the switch

In September 2022, six Ukrainian drone submarines packed with explosives were navigating through the dark waters off the Crimean Peninsula toward the Russian Black Sea Fleet anchored at Sevastopol. The drones were guided by Starlink — the satellite internet constellation operated by SpaceX, which is owned by Elon Musk. Ukrainian military planners had designed the operation to strike the fleet that had been bombarding Ukrainian coastal cities and enforcing a naval blockade for seven months.

The drones lost connectivity. They washed ashore harmlessly. The Russian fleet survived.

Musk had refused to extend Starlink coverage to Crimea. He later stated that activating Starlink around Sevastopol would have made SpaceX complicit in a major act of war. He had spoken with the Russian ambassador to the United States, who had told him that a Ukrainian attack on Crimea could provoke a nuclear response. The decision to deny coverage was made by one person. Not a head of state. Not a military commander. Not an elected official of any government. A private citizen who owned the infrastructure.

The Ukrainian perspective on the decision was captured by a participant in the operation, who recalled the moment communication was cut: the team was seventy kilometers from the target, the operation was underway, and the connectivity disappeared. Ukrainian officials attempted to persuade Musk to restore coverage. American officials were contacted. The response from the Americans: it was a private company, and they could not put pressure on it.

Mykhailo Podolyak, a senior advisor to President Zelenskyy, summarized the consequence: by preventing the Ukrainian drone attack, the decision allowed the Russian fleet to continue firing Kalibr cruise missiles at Ukrainian cities. The fleet that was not sunk that night continued to bombard civilian infrastructure for months afterward.

The Switch

Three and a half years later, in February 2026, the switch flipped in the other direction.

Evidence had emerged that Russian forces were using Starlink terminals — obtained through third-party countries — to guide long-range strike drones into Ukrainian cities. A Russian BM-35 drone had penetrated central Kyiv, flying so low that officials in the Cabinet of Ministers building could watch it pass beneath them from the seventh floor. Ukrainian Defense Minister Mykhailo Fedorov presented Musk with evidence of the Russian military's use of his network.

Musk cut off unauthorized Russian access to Starlink.

The effect was immediate and devastating. Within five days, Ukrainian forces reclaimed over two hundred square kilometers of territory — roughly equivalent to Russia's gains for the entire previous month. Russian frontline communications collapsed. Units could not coordinate. In one incident on the Zaporizhzhia front, twelve Russian soldiers were killed by friendly fire after a Starlink terminal failure disrupted their coordination. Russia's own satellite communications system, operated by Gazprom Space Systems, was regarded as far less reliable and could not fill the gap.

Ukrainian cyber operatives, exploiting the chaos, built a fake Starlink registration portal and invited Russian soldiers to register terminals for a fee. The scam worked: Russia's 256th Cyber Assault Division collected the precise locations of 2,420 enemy Starlink terminals in a single week — data that was fed directly to artillery and drone targeting units.

Two hundred square kilometers. Twelve soldiers killed by friendly fire. 2,420 terminals exposed. Because one person flipped a switch on a commercial satellite network. Not a weapon. Not a military system. A commercial internet service that both sides of a war had incorporated into their military operations because no alternative of comparable capability existed.

The Dependency

The European response to the February 2026 events was not celebration. It was alarm.

The alarm was not about the decision itself but about the dependency it revealed. European defense officials recognized that the same system that cut Russia off in February could theoretically cut Ukraine off in March. The switch belonged to one person. The governance belonged to no one. As one European defense industry executive stated: Europe cannot build a military strategy that runs through one man's decisions.

Starlink operates approximately seven thousand satellites — over sixty percent of all active satellites in low Earth orbit. In Ukraine, over 42,000 Starlink terminals are used by the military, hospitals, businesses, and government agencies. During Russian bombing campaigns that caused widespread blackouts, Ukraine's public agencies turned to Starlink to stay online. The dependency is not incidental. It is structural. No alternative of comparable capability, coverage, and resilience exists. Europe's own satellite communications programs are years from operational deployment at equivalent scale.

The US Air Force Secretary, responding to the original Crimea incident, articulated the problem in terms that apply to every domain this chapter examines: "If we're going to rely upon commercial architectures or commercial systems for operational use, then we have to have some assurances that they're going to be available. Otherwise they are a convenience and maybe an economy in peacetime, but they're not something we can rely upon in wartime."

The assurances do not exist. No governance framework governs the obligations of a private satellite constellation operator during armed conflict. No treaty addresses the scenario in which both sides of a war depend on the same commercial infrastructure controlled by a single private individual. No international body has jurisdiction over orbital communications monopolies. No framework exists. And a war is being fought on infrastructure that belongs to a person who is neither a party to the conflict nor accountable to any of the governments whose forces depend on his network.

The honest accounting requires acknowledging the other side. No government, no international consortium, no distributed democratic process deployed seven thousand satellites in low Earth orbit. One company did. The connectivity Ukraine depended on in 2022 existed because of the concentration of capability this chapter warns about. A distributed, democratically governed satellite network would have taken a decade to build and would not have been available when Ukraine needed it. The tradeoff is real: concentration enables speed; distribution enables governance. The question is not whether concentration has benefits — it manifestly does. The question is whether the benefits justify a structure in which one person's decisions determine the trajectory of a war.

The Compute Bottleneck

Starlink is the most dramatic example of infrastructure consolidation, but it is not the most consequential for the governance of artificial intelligence. That distinction belongs to the compute bottleneck.

Training a frontier AI model — a system capable of the broad, general-purpose reasoning demonstrated by the most advanced chatbots, code generators, and analytical tools — requires three things: vast quantities of training data, specialized hardware (graphics processing units or tensor processing units, configured in clusters of thousands), and months of continuous computation. The cost of

training a single frontier model is measured in hundreds of millions of dollars. The number of organizations on Earth with the financial resources, technical expertise, and hardware access to train a frontier model is approximately seven: OpenAI, Anthropic, Google DeepMind, Meta, Mistral, xAI, and a small number of Chinese labs including Baidu and Alibaba's research divisions.

Seven organizations. In a world of eight billion people, 195 nations, and millions of institutions, the capacity to produce the most consequential technology of the twenty-first century is concentrated in fewer than ten organizations, most of them headquartered within a fifty-mile radius of each other in the San Francisco Bay Area.

The concentration deepens at each layer of the stack. The specialized chips required for AI training are designed primarily by one company: NVIDIA. They are manufactured primarily by one company: TSMC, based in Taiwan. The cloud infrastructure on which most AI workloads run is provided by three companies: Amazon Web Services, Microsoft Azure, and Google Cloud. The pipeline from chip design to chip fabrication to cloud deployment to model training to model serving passes through a remarkably small number of chokepoints — narrow control points in the supply chain where a single entity determines what passes through, each controlled by a single entity or a small oligopoly.

The Pattern Recognized

The architecture is the one traced across five thousand years in *Who Holds the Jar?*.

The mainframe era concentrated computing power in a handful of companies — IBM above all. The personal computer distributed capability to millions — and then Microsoft and Intel recaptured the value through control of the operating system and the processor. The internet distributed communication to billions — and then five companies recaptured the infrastructure. The cycle repeats: distribute the access, recapture the control. Each era feels like democratization. Each era produces reconcentration.

AI is following the pattern with a new feature: the reconcentration is happening *simultaneously* with the democratization. ChatGPT was released to the public in November 2022. By the time it reached 100 million users two months later, the companies that built it had already secured the compute, the data, and the distribution channels that would ensure the value flowed to them. The previous cycles took decades to move from democratization to reconcentration. AI is doing it in real time. The users experience the distribution. The companies capture the value. The two happen concurrently, and the gap between them is closing before any governance framework can pry it open.

The Stack

To understand the full scope of the consolidation, it helps to visualize the AI infrastructure as a vertical stack — a series of layers, each built on the one below it, each controlled by a small number of entities.

At the bottom: the chips. NVIDIA designs the GPUs that train AI models. TSMC manufactures them. The design and fabrication of the most advanced AI chips are concentrated in two companies, separated by the Pacific Ocean, in a geopolitical relationship (US-Taiwan-China) that is among the most volatile on earth. A disruption to either company — through conflict, natural disaster, or sanctions — would constrain the global AI industry at its foundation.

Above the chips: the compute. The chips are assembled into data centers operated by cloud providers — primarily AWS, Azure, and Google Cloud. These three companies control the majority of the world's AI-capable compute infrastructure. A company or government that wants to train an AI model but does not own a data center must rent compute from one of these three providers. The providers set the prices, the terms of service, and the acceptable use policies. They are, in effect, the landlords of the AI economy.

Above the compute: the models. The frontier AI models are built by approximately seven organizations. These models are the engines that power every downstream application — from chatbots to code generators to diagnostic tools to targeting systems. The organizations that build them determine their capabilities, their limitations, their safety properties, and — through acceptable use policies and content filters — their behavior. The model builders are not just technology companies. They are governance actors, whether they acknowledge it or not, because the decisions they make about what their models can and cannot do are decisions that affect every application built on top of them.

Above the models: the distribution. The models reach users through app stores, API access, enterprise agreements, and integration into existing products (Microsoft's integration of OpenAI's models into Office, Google's integration of Gemini into Search). The distribution channels are controlled by the same companies that control the compute and, in some cases, the models themselves. The vertical integration is remarkable: a single company can own the cloud infrastructure, build the model, and distribute it through its own products.

And wrapping around everything: the communications layer. The orbital infrastructure that connects the user to the service. Sixty percent of active satellites in low Earth orbit belong to one company. The physical layer on which everything else depends — the ability to reach the cloud, access the model, use the application — passes through a chokepoint controlled by a single individual whose decisions, as this chapter opened by documenting, can determine the outcome of a war.

The Governance Question

The consolidation of AI infrastructure is not a market failure in the traditional sense. It is a structural consequence of the technology's requirements. Training frontier models requires enormous capital investment, specialized hardware, and concentrated expertise. The barriers to entry are not artificial — they are the physics and economics of the technology itself. No antitrust action can make frontier AI training cheap.

But the concentration of capability in a small number of entities creates a governance problem that no existing framework addresses. The entities that control the AI stack are not regulated utilities. They are not common carriers. They are not subject to the obligations that apply to companies whose infrastructure is deemed essential to the public interest. They set their own terms. They determine their own acceptable use policies. They decide, without democratic input, what their models can and cannot do.

When Samuel Insull's utility empire collapsed in the 1930s — as described in Chapter 2 — the governance response was the Public Utility Holding Company Act: regulation that recognized electrical infrastructure as essential and imposed obligations commensurate with that status. The AI infrastructure is essential. The cloud compute that runs AI workloads also runs healthcare systems, financial markets, government services, and critical infrastructure. The orbital communications layer connects military operations, hospitals, and governments in conflict zones. The models themselves are being embedded in medical diagnosis, legal research, educational instruction, and military targeting.

The infrastructure is essential. The governance treats it as optional. The gap between the two is the gap this book exists to name.

"The infrastructure is essential. The cloud runs healthcare, finance, and government. The orbital layer connects military operations and hospitals. The

The Two Gatekeepers

This chapter has described two forms of gatekeeping, and naming them clearly is essential for the chapters that follow.

The first is infrastructure gatekeeping: control of the physical and computational resources on which everything else depends. TSMC's chip fabrication. NVIDIA's chip design. The cloud providers' data centers. Starlink's orbital constellation. These gatekeepers control access to the substrate. Without their resources, AI does not function. Without their network, the drone does not reach its target and the hospital does not stay online during a blackout. Infrastructure gatekeeping is control through exclusion: the power to deny access.

The second is capability gatekeeping: control of what the AI can do. The model builders who determine their systems' safety properties, content policies, and acceptable use restrictions. Anthropic's

decision that its models will not be used for autonomous weapons or mass surveillance — traced in *The Last Keeper*, — is an act of capability gatekeeping. So is every content filter, every refusal policy, every safety benchmark. Capability gatekeeping is control through design: the power to shape what the technology does.

Both forms of gatekeeping are currently exercised by private companies, at their discretion, without democratic mandate. Musk's Starlink decisions are infrastructure gatekeeping: the power to grant or deny connectivity that determines the outcome of military operations. Anthropic's red lines are capability gatekeeping: the power to determine which uses of AI are permissible. Both decisions may be defensible on their merits. Both decisions were made by private actors exercising judgment that, in a governed world, would be exercised by institutions with democratic accountability.

The question is not whether these gatekeepers are making the right decisions. The question is whether decisions of this magnitude — decisions that determine the viability of states, the outcome of wars, the boundaries of AI capability — should be made by private actors at all, regardless of how wise or well-intentioned those actors are. The lighthouse keeper at Smalls was one person watching one light. The new gatekeepers are private individuals and companies holding switches that affect billions of lives. The scale demands governance. The governance does not exist.

The consolidation is domestic and global simultaneously. The next chapter widens the lens to the geopolitical dimension: the race between nations to control AI capability, the semiconductor export controls that weaponized the compute bottleneck, and the nuclear parallel that provides the only historical precedent for governing a technology of this magnitude at international scale.

The Great Convergence

The Race

On the geopolitical competition for AI supremacy, the weaponization of compute, and the question of whether democracy can compete

"The race is not between nations. It is between the pace of capability and the pace of governance. And capability is winning."

The Great Convergence

The Race

On the US-China AI competition, the semiconductor export controls that weaponized the supply chain, and the question of whether AI governance can be achieved at AI speed rather than nuclear speed

"The governance is national. The technology is planetary. No nation can govern AI alone. No two nations agree on how to govern it together."

The Race

On the geopolitical competition for AI supremacy, the weaponization of the supply chain, and the nuclear parallel

On October 7, 2022, the United States Bureau of Industry and Security published a set of export controls that would reshape the global AI landscape. The controls restricted the sale of advanced semiconductor chips and chip-manufacturing equipment to China. The restrictions were not targeted at a specific company or a specific product. They were targeted at a capability: China's ability to train frontier AI models.

The controls leveraged the chokepoints previous chapter. NVIDIA's most advanced AI chips — the A100 and H100 — could no longer be sold to Chinese customers. The chip-manufacturing equipment produced by ASML, the Dutch company that makes the extreme ultraviolet lithography machines essential for fabricating the most advanced chips, was also restricted through coordinated agreements with the Netherlands and Japan. The intent was to deny China access not just to the chips themselves but to the machinery needed to build them.

The United States was using its position at the top of the semiconductor stack — the same stack described in Chapter 7 — as a geopolitical weapon. The compute bottleneck was not merely a market outcome. It was now a tool of strategic competition. The country that controlled the chokepoints could determine who had access to the most consequential technology of the twenty-first century and who did not.

"The semiconductor export controls were not targeted at a product. They were targeted at a capability: China's ability to train frontier AI models. The compute bottleneck became a geopolitical weapon."

Two Philosophies

The US-China AI competition is often framed as a technology race — which nation will build the more capable AI system first. This framing is not wrong, but it is incomplete. The competition is also, and more fundamentally, a governance race: two superpowers with incompatible visions of how AI should be governed, each building the technology according to its own philosophy, each attempting to ensure that its philosophy prevails.

The Chinese approach integrates AI development with state objectives. China's AI strategy, articulated in the New Generation Artificial Intelligence Development Plan of 2017, envisions AI as a tool of national power: economic competitiveness, military capability, and social management. The surveillance infrastructure — the Xinjiang monitoring system, the social credit framework, the integrated surveillance platforms — is not a side effect of China's AI described in The Last Keeper

described in The Last Keeper

development. It is the development. The technology and its governance are fused: the state builds the AI, deploys the AI, and determines the AI's purpose. The governance is total but not democratic. The rules exist. They are written by the state, for the state.

The American approach separates AI development from state control — at least in principle. The United States' AI leadership resides primarily in private companies, funded by private capital, competing in private markets. The government's role is regulatory rather than directive: setting guardrails (or attempting to) through executive orders, legislation, and agency oversight. The private companies determine what the AI can do. The government determines — or fails to determine — what the AI may do. The separation produces innovation. It also produces the governance vacuum that seven chapters.

Neither philosophy is adequate on its own. The Chinese model produces governance without liberty — rules written by the state, enforced by the state, serving the state's interests rather than the citizen's. The American model produces liberty without governance — innovation driven by private actors who operate faster than any regulatory framework can follow, in a political system where the lobbying power of the technology companies exceeds the regulatory capacity of the agencies tasked with governing them.

> *"The Chinese model produces governance without liberty. The American model produces liberty without governance. Neither is adequate. The world needs both."*

The Response

China's response to the semiconductor export controls was not capitulation. It was acceleration.

Chinese AI labs, denied access to the most advanced NVIDIA chips, adapted their training methods to work with less capable hardware. They developed techniques for training models more efficiently, squeezing more capability from fewer resources. Huawei accelerated the development of its Ascend series of AI chips — less powerful than NVIDIA's best but capable of training competitive models. Chinese companies stockpiled chips purchased before the controls took effect and obtained restricted chips through intermediary countries that were not subject to the export restrictions.

The controls did not stop Chinese AI development. They may have slowed it. They certainly changed its trajectory — pushing Chinese labs toward efficiency innovations that American labs, with abundant compute, had less incentive to pursue. And they accelerated China's drive toward semiconductor self-sufficiency — a program that, if successful, would eliminate the chokepoint the controls were designed to exploit.

The dynamic is familiar to anyone who has studied arms control. Restrictions on a capability create incentives to develop alternative paths to the same capability. The restriction buys time. It does not solve the problem. The problem is that two nations with incompatible governance philosophies are both building documented across

the most powerful technology in human history, and no framework exists for governing the competition itself.

The Fragmented Landscape

The United States and China are the dominant competitors, but they are not the only actors. The global AI governance landscape is a patchwork of national and regional approaches, each reflecting different priorities, different risk assessments, and different political constraints.

The European Union enacted the AI Act in 2024 — the first comprehensive AI regulatory framework in the world. The Act classifies AI systems by risk level and imposes requirements proportional to the risk: minimal regulation for low-risk applications, strict requirements for high-risk systems used in areas like healthcare, law enforcement, and critical infrastructure, and outright prohibition of certain uses deemed incompatible with fundamental rights, including real-time remote biometric identification in public spaces for law enforcement purposes. The EU's approach is the most systematic attempt at AI governance to date. It is also, by the admission of its proponents, a framework designed for the European context — one jurisdiction's rules for a technology that does not respect jurisdictional boundaries.

The United Kingdom has pursued a "pro-innovation" approach — lighter regulation, with sector-specific guidance rather than horizontal legislation. The philosophy is that heavy regulation would drive AI companies and investment to less regulated jurisdictions. Canada, Japan, South Korea, Singapore, and other nations have published AI strategies, guidelines, and in some cases legislation, each calibrated to domestic political conditions and economic ambitions.

The fragmentation is the problem. AI models are trained in one jurisdiction, hosted in another, accessed from a third, and deployed in all of them simultaneously. A model trained in the United States, hosted on servers in Ireland, accessed by users in Brazil, and deployed in a healthcare system in Nigeria is subject to — in theory — the regulations of every jurisdiction it touches. In practice, it is subject to the regulations of whichever jurisdiction has the least enforcement capacity, because the companies that build and deploy the models can route around restrictive regulation by hosting in permissive jurisdictions. The governance is national. The technology is planetary.

> *"A model trained in the US, hosted in Ireland, accessed from Brazil, deployed in Nigeria is subject to every jurisdiction it touches. In practice, it is subject to the jurisdiction with the least enforcement. The governance is national. The technology is planetary."*

The Rupture

On January 20, 2026, Canadian Prime Minister Mark Carney stood before the World Economic Forum in Davos and delivered a speech that received a rare standing ovation from the assembled heads of state

and business leaders. The speech was, in the assessment of multiple international analysts, the most significant statement by a Western leader on the collapse of the post-Cold War order since the order began to fracture.

Carney's argument was built on a literary reference — Václav Havel's description of how citizens in authoritarian states sustain the system by participating in rituals they privately know to be false. Carney applied Havel's framework to the rules-based international order: the system's power came not from its truth but from everyone's willingness to perform as if it were true. Canada and other nations had willingly participated in an order they knew to be partially false — overlooking the unevenness of its application — because they benefited from its predictability. American hegemony had provided genuine public goods: open sea lanes, a stable financial system, collective security, and frameworks for resolving disputes.

That bargain, Carney declared, no longer works. His formulation was precise and unsparing: "We are in the midst of a rupture, not a transition." Great powers had begun using economic integration as weapons — tariffs as leverage, financial infrastructure as coercion, supply chains as vulnerabilities to be exploited. The integration that had been sold as mutual benefit had become a source of subordination. Nations that negotiated bilaterally with a hegemon negotiated from weakness. They accepted what was offered. They competed with each other to be the most accommodating. "This is not sovereignty," Carney said. "It's the performance of sovereignty while accepting subordination."

The passage most relevant to this book — and to the governance challenge it documents — was Carney's statement on artificial intelligence. In three sentences, he named the exact convergence that Chapters 7 and 8 of this book have been tracing: "On AI, we are cooperating with like-minded democracies to ensure we will not ultimately be forced to choose between hegemons and hyperscalers. It is building coalitions that work."

Hegemons and hyperscalers. Two words that name the two forms of gatekeeping described in Chapter 7: the geopolitical powers that control the strategic environment, and the technology companies that control the infrastructure. Carney's formulation acknowledged what the fragmented governance landscape obscures: that the choice facing nations is not between different regulatory approaches but between different forms of dependence. Dependence on a hegemon whose interests may not align with yours. Or dependence on hyperscalers whose infrastructure you cannot replicate and whose terms you cannot negotiate. The satellite constellation that controls sixty percent of low Earth orbit. The chip designer. The chip manufacturer. The cloud providers. The model builders. The stack described in the previous chapter, restated as a geopolitical reality by a sitting head of state. Not naive multilateralism.

Carney's solution was coalitions — middle powers combining to create what he called "a third path with impact." Collective investments in resilience, shared standards to reduce fragmentation, buyer's clubs to diversify supply of critical minerals, and cooperation among like-minded democracies on AI governance. The approach was neither the American model (liberty without governance) nor the Chinese model (governance without liberty). It was an attempt to build governance from the middle — from the nations that have the most to lose from a world in which the rules are written exclusively by the powerful.

His closing formulation landed with the force of a principle: "Middle powers must act together, because if you're not at the table, you're on the menu." The argument applied not only to nations. The patients affected by biased healthcare algorithms are not at the table. The defendants sentenced by proprietary risk scores are not at the table. The workers displaced by AI systems are not at the table. The surveilled communities are not at the table. Carney was speaking about nations, but the principle is universal: absent from the process, present for the consequences. The missing seats are not only national. They are individual. And the cost of absence is borne by whoever is not in the room.

Critics have argued that Carney's simultaneous engagement with Beijing — lowering tariffs on Chinese electric vehicles in exchange for agricultural access — undermines the sovereignty argument by substituting one dependence for another. The objection is not frivolous. But the distinction between bilateral negotiation on specific sectors and structural dependence on another power's infrastructure is precisely the distinction this chapter has been drawing. A trade deal is a policy choice. A satellite constellation you do not own is a dependency you cannot exit.

The Nuclear Parallel

The closest historical parallel to the current AI governance challenge is the governance of nuclear weapons — the case study from Chapter 2 that demonstrated the fastest progression from innovation to governance in the history of transformative technology.

The parallels are substantial. Nuclear weapons emerged from a national security project during wartime. AI emerged from research labs during a period of intensifying geopolitical competition. Nuclear capability was initially concentrated in one nation and then proliferated to rivals. AI capability is concentrated in a handful of companies in two nations and is proliferating through open-source models, espionage, and parallel development. Nuclear weapons posed an existential risk that no single nation could manage alone. AI poses risks — to labor, to information integrity, to military stability, to surveillance and civil liberties — that no single nation can govern alone.

The NPT, signed in 1968, provided a framework — imperfect, inequitable, but functional — that has prevented the proliferation of nuclear weapons to dozens of nations that had the technical capacity to build them. It took 23 years from Trinity to the NPT. During those 23 years, the world came within hours of nuclear war during the Cuban Missile Crisis, atmospheric testing contaminated the planet, and the populations of Hiroshima and Nagasaki provided the evidence of what the technology could do when used. The governance arrived after two cities were destroyed and the world nearly followed.

The question for AI governance is whether the nuclear pattern can be compressed — whether governance can arrive before the equivalent of Hiroshima rather than after it. The nuclear precedent provides both the model (international treaty, verification mechanisms, institutional enforcement) and the warning (the governance required catastrophe as a prerequisite). The question is which aspect of the precedent will repeat.

Why AI Is Harder

The nuclear parallel illuminates the governance challenge, but it also reveals critical differences that make AI governance harder than nuclear governance in several respects.

The technology is dual-use — capable of both civilian and military application — at every level. Nuclear weapons and nuclear power can be distinguished — imperfectly, but the distinction exists in the physical infrastructure required. Enrichment facilities produce observable signatures. Weapons-grade material has specific properties. The IAEA can inspect. AI has no equivalent distinction. The same model that writes a student's essay can generate disinformation. The same training infrastructure that produces a medical diagnostic tool can produce a military targeting system. The same code generation tool that helps Maya build a patient management system can help a malicious actor build malware. There is no physical marker that distinguishes a beneficial AI system from a harmful one. The capability is identical. The intent is invisible.

The actors are not only states. Nuclear weapons are built by governments. AI is built by companies. The NPT is a treaty between nations. A treaty between nations cannot directly govern companies whose operations span multiple jurisdictions and whose compliance depends on corporate willingness rather than sovereign obligation. The Anthropic red lines, the OpenAI safety commitments, the Google DeepMind ethics board — these are voluntary corporate governance mechanisms. They can be changed by a board vote. They can be abandoned under competitive pressure. They are not treaties.

The proliferation is inherent. Nuclear nonproliferation works, imperfectly, because the materials and facilities required to build a nuclear weapon are difficult to acquire, expensive to maintain, and detectable

by inspection. AI proliferation is inherent in the technology's nature. Open-source models can be downloaded by anyone. The knowledge to train models is published in academic papers. The hardware, while expensive for frontier models, is sufficient for capable smaller models at consumer prices. You cannot prevent the proliferation of AI the way you can prevent the proliferation of enriched uranium. The knowledge is in the open. The genie is out of the bottle.

The speed exceeds governance capacity. The NPT took 23 years to negotiate. AI capabilities are advancing on a timeline measured in months. A governance framework negotiated through traditional diplomatic channels — the pace of the UN, the speed of treaty ratification, the tempo of multilateral negotiation — will be obsolete before it is signed. The technology will have evolved past whatever the framework was designed to govern.

What Governance Would Require

Despite these difficulties, the nuclear precedent provides elements that any AI governance framework would need to include.

A shared assessment of risk. The NPT was possible because the major powers agreed on one thing: uncontrolled nuclear proliferation was a shared threat. The Cuban Missile Crisis provided the catalytic moment — the event that made the risk undeniable. AI governance requires a similar shared assessment. Some elements of that assessment are emerging — the Bletchley Declaration of

November 2023, signed by 28 countries including the US and China, acknowledged the risks of frontier AI. But acknowledgment is not governance. A declaration is not a treaty.

Verification mechanisms. The International Atomic Energy Agency — the IAEA — provides the inspection and verification infrastructure for the NPT. No equivalent institution exists for AI. Proposals for international AI safety institutes, model registries, and compute monitoring frameworks are under discussion but not yet operational at scale. The technical challenge of verifying AI development is harder than verifying nuclear development: a data center training an AI model looks identical from the outside to a data center running a cloud computing workload.

Institutional enforcement. The IAEA has the authority to inspect facilities, report violations, and refer non-compliance to the UN Security Council. No AI governance institution has equivalent authority. The EU AI Act has enforcement mechanisms within Europe. No institution has enforcement authority at the global level — the level at which the technology operates.

A governance speed that matches the technology speed. that has no precedent. Every previous governance framework was written for a technology that had stabilized. The printing press did not evolve between Gutenberg and the Statute of Anne. Nuclear weapons advanced, but the fundamental physics was understood. AI is evolving faster than any previous governed technology. A framework designed for today's AI may not address tomorrow's. The governance must be adaptive — designed not for a specific capability but for a class of capabilities that is changing as the rules are being The requirement

written.

> *" technology that had stabilized. AI is evolving faster than any previous governed technology. The governance must be adaptive — designed not for a specific capability but for a class of capabilities that is changing while the rules are being written."*

Part III Restated

Part III has mapped the consolidation phase of AI's five-phase cycle across two dimensions.

Chapter 7 mapped the domestic consolidation: the concentration of AI infrastructure in a small number of companies, the vertical integration of the stack from chip to cloud to model to distribution, the orbital communications monopoly that wraps around everything, and the two forms of gatekeeping — infrastructure and capability — that are exercised by private actors without democratic mandate.

This chapter mapped the geopolitical consolidation: the competition between two superpowers with incompatible governance philosophies, the weaponization of the supply chain through export controls, the fragmented global regulatory landscape that cannot govern a planetary technology with national rules, and the nuclear parallel that provides both a model and a warning for what international AI governance might look like.

The consolidation is proceeding on both dimensions simultaneously. Domestically, a handful of companies control the infrastructure. Geopolitically, two nations are competing to control the capability. The governance that could address either dimension — domestic regulation of essential AI infrastructure, or international coordination of AI development — does not yet exist in a form adequate to the challenge.

Part IV enters the room. Who is attempting to write the rules? What dilemmas do they face? Who is absent from the process? And what happens when the three trajectories of this trilogy — power concentration, the removal of the keeper, and the absence of governance — converge?

Prior governance frameworks were written for a

The Great Convergence

The Builders' Dilemma

On the companies that build the technology and the impossible choice they face between profit and principle

"An AI company drew two lines: no mass surveillance of citizens, no autonomous weapons without human oversight. For this, it was designated a supply-chain risk threat."

The Builders' Dilemma

On the companies that build the technology and the impossible choice they face between self-governance and survival

In February 2026, Anthropic — the company that builds the Claude AI system — was offered a contract with the United States Department of Defense valued at approximately $200 million. The company's leadership evaluated the contract and identified two applications that it determined were incompatible with its values: the use of its AI for autonomous weapons targeting and for mass surveillance of civilian populations.

Anthropic declined those two applications while remaining open to other defense work. The company did not refuse the relationship. It drew lines within the relationship — two specific boundaries around the uses it would not permit.

The response from the defense establishment was swift and severe. Elements within the Department of Defense pushed to designate Anthropic as a supply-chain risk — a classification that would effectively blacklist the company from all government contracts, not just the disputed ones. The argument was not that Anthropic's technology was inadequate. The argument was that a company unwilling to provide unrestricted AI capabilities to the military could not be relied upon as a strategic partner. The message was clear: if you build the technology, you do not get to decide how it is used. The customer decides. And this customer has a budget measured in hundreds of billions of dollars.

The public response was different. Within weeks of the confrontation becoming public, Claude became the number one application in the App Store. The company reported over one million daily sign-ups. The growth was driven predominantly by new users — people who had never previously used the product, who chose Anthropic specifically because the company had drawn a line. The market rewarded the refusal.

The structural tension that the Anthropic confrontation made visible — a tension that exists inside every company that builds frontier AI, and that no company can resolve alone.

> *"The defense establishment said: if you build the technology, you do not get to*
> *decide how it is used. The public said: we will choose the company that*
> *decides. Both responses were rational. They pointed in opposite directions."*

Five days before the Anthropic confrontation became public, an identical dynamic played out at a different scale. Canadian Prime Minister Mark Carney stood before the World Economic Forum in Davos and named the rupture in the rules-based international order — the speech documented in Chapter 8 of this book. The response from the hegemon followed the same architecture as the response to Anthropic's red lines. Carney was revoked from the Board of Peace. He was threatened with 100% tariffs. He was called "Governor Carney" — a deliberate diminishment of sovereignty. The US Treasury Secretary described the oil-rich province of Alberta as "a natural partner for the US" while meeting with separatist groups seeking to fragment Canada from within. And when Carney refused to retract, the administration claimed he was "aggressively walking back" his remarks — a claim Carney denied publicly and on the record: "I meant what I said in Davos."

Anthropic drew a line and was threatened with blacklisting. Carney named a reality and was threatened with economic devastation. pattern is the same: the entity that self-governs — that exercises judgment about what is permissible — is punished by the entity with power. The dilemma is not confined to companies. It operates at every scale where an actor with less power attempts to constrain an actor with more.

The Dilemma Stated

The builders' dilemma is a coordination problem, and stating it precisely matters because the precision reveals why market solutions are insufficient.

Scenario one: unilateral self-governance. A company governs itself — draws red lines, restricts certain uses, invests in safety research, declines contracts that cross its boundaries. If its competitors do not do the same, the self-governing company loses market share, government contracts, and potentially its viability. The customers who want the unrestricted capability go to the competitors who provide it. The self-governance is punished by the market. The company either abandons its lines or is outcompeted by companies that never drew them.

Scenario two: universal self-governance. Every company in the industry governs itself — adopts the same red lines, restricts the same uses, invests at the same level in safety. No company loses competitive advantage because all companies bear the same constraints. The problem is solved. But this scenario requires simultaneous, voluntary, sustained coordination among competitors — companies that are, by definition, trying to outperform each other. Voluntary coordination of this kind is, in the language of game theory, an unstable equilibrium. The incentive to defect — to quietly relax one's constraints while competitors maintain theirs — is persistent and structural. One defection breaks the equilibrium. And the competitive pressure is existential.

Scenario three: no self-governance. No company governs itself. All capabilities are deployed without restriction. The chaos phase deepens. The harms examined in Parts I and II of this book — and in the two books that precede it — accelerate. Eventually, the harms produce a crisis severe enough to trigger reactive government regulation. Reactive regulation, designed in crisis, is consistently worse than proactive governance: it is broader, blunter, less technically informed, and more likely to constrain beneficial uses along with harmful ones. The Sarbanes-Oxley Act, written in the aftermath of the Enron scandal, imposed compliance costs on every public company to prevent the behavior of a few. The Patriot Act, written in the aftermath of September 11, expanded surveillance authorities that have proven extremely difficult to roll back. Reactive regulation is the worst of all outcomes — and it is the outcome that

The coercion extended beyond rhetoric. In January and February 2026, the United States aggressively pursued the annexation of Greenland — an autonomous territory of Denmark, a founding NATO ally. When European nations attempted to bolster Arctic security in response to what they interpreted as a direct territorial threat to a European ally, the administration threatened a 10% tariff on those nations. The economic weapon was deployed not against an adversary but against allies who attempted to uphold the collective security framework that the United States itself had built.

The Greenland annexation threats, the Alberta separatist meetings, the tariff threats, the deliberate diminishment of an allied head of state — these are not isolated incidents. They are the instruments of a power that has concluded that the rules-based order it built is now a constraint on its freedom of action. The hegemon is not abandoning the room. It is rebuilding the room with fewer seats, higher walls, and a lock on the inside. The Board of Peace — established at the World Economic Forum in January 2026 to handle Gaza's postwar reconstruction, deliberately bypassing the UN Security Council, with the US as the only permanent UNSC member at the table — is the architectural drawing of the new room. It is a governance structure designed by a hegemon to exclude the traditional multilateral checks and balances that the hegemon itself established eighty years ago at Bretton Woods.

no self-governance produces.

> *"Self-governance alone is punished by the market. Coordinated*
> *self-governance is an unstable equilibrium. No self-governance produces*
> *reactive regulation — the worst outcome. The dilemma has no market solution.*
> *That is why it requires a room."*

The Landscape of Attempts

The Anthropic confrontation was the most visible instance of the dilemma, but it was not the first. The AI industry's recent history is a catalog of attempts to self-govern — and the structural forces that undermine those attempts.

In 2018, Google established an AI ethics board — the Advanced Technology External Advisory Council. It was disbanded within a week of its announcement, after controversy over the appointment of members whose views on climate change and LGBTQ rights were deemed incompatible with the board's mission. The attempt to create an independent governance mechanism for Google's AI development collapsed before it held a single meeting.

In 2020, Google fired Timnit Gebru — the co-author of the Gender Shades research and one of the most prominent AI ethics researchers in the world — after a dispute over a paper that examined the risks of large language models. The departure of Gebru, and the subsequent departure of her co-lead Margaret Mitchell, effectively dismantled Google's internal AI ethics team. The message was legible: internal governance that conflicts with the company's product direction does not survive.

OpenAI's evolution traced a different path to the same conclusion. Founded in 2015 as a nonprofit dedicated to ensuring that artificial general intelligence benefits humanity, OpenAI restructured in 2019 to a "capped-profit" model, then progressively moved toward a for-profit structure. The original mission — to develop AGI safely and distribute its benefits broadly — was not abandoned in rhetoric. But the structural incentives shifted: the company took billions in investment from Microsoft, became a commercial product company, and the safety mission was now competing for resources and attention with the revenue mission. Several senior safety researchers departed, citing concerns that safety work was being deprioritized relative to product development.

Meta took yet another approach: open-source release of its Llama models. The philosophy was that open access would democratize AI capability and prevent the concentration of power in a small number of closed-model companies. The governance implication, however, was that open-sourced models are, by definition, ungoverned after release. Once the model weights are public, anyone can fine-tune the model for any purpose, remove safety guardrails, and deploy the result without the original developer's knowledge or consent. Meta's open-source strategy was a genuine contribution to the democratization of traced in The Last Keeperand

AI capability. It was also, simultaneously, a contribution to the governance vacuum.

This trilogy must be honest about the paradox. *Who Holds the Jar?* traces five thousand years of computational power concentrating in fewer and fewer hands and identifies that concentration as the existential danger. Open-source AI is the most powerful mechanism available for breaking that concentration — distributing capability to anyone who wants it. This trilogy warns against concentration and warns against ungoverned distribution. Both warnings are correct. And they pull in opposite directions. The honest answer is that neither pure concentration nor pure distribution is safe. What is needed is distribution with standards — open access to capability, within frameworks that establish floors below which no deployment may fall. AI presents, and the fact that it is hard is not an argument for ignoring it.

The Structural Tension

The pattern across these cases is consistent and instructive. Every major AI company has, at some point, attempted to govern its own technology. Google created an ethics board. OpenAI was founded as a nonprofit. Anthropic drew red lines. Meta open-sourced its models in the name of democratization.

And was constrained, compromised, or undermined by the same force: competitive pressure. Google's ethics board was disbanded because its existence created controversy that threatened the company's reputation. Google's ethics team was dismantled because its findings conflicted with product direction. OpenAI's nonprofit structure was abandoned because the capital requirements of frontier AI development exceeded what a nonprofit could raise. Meta's open-source releases advanced democratization but precluded downstream governance. And Anthropic's red lines, however principled, exist at the pleasure of the company's leadership — they can be changed by a board vote, under competitive pressure, without external review.

The structural tension is this: building frontier AI requires billions of dollars in capital. Capital comes from investors who expect returns. Returns require revenue. Revenue requires customers. Customers include governments and enterprises that want unrestricted capability. Restricting capability restricts the customer base. Restricting the customer base restricts revenue. Restricting revenue threatens the capital. And threatening the capital threatens the company's ability to build frontier AI at all.

The chain from safety commitment to existential risk is direct: a company that governs too aggressively may not survive to govern at all. It is the explicit logic behind the supply-chain risk designation that was threatened against Anthropic. The message from the defense establishment was: your self-governance is a risk to our supply chain. Your principled refusal is our strategic vulnerability. Govern yourself less, or be excluded from the market that funds your existence.

"The chain from safety commitment to existential risk is direct: a company that governs too aggressively may not survive to govern at all. It

each time, the self-governance

cannot be resolved from inside the market."

Why Markets Cannot Solve It

The builders' dilemma is, in its formal structure, a multi-player prisoner's dilemma — a well-studied problem in game theory with a well-known result: in the absence of an enforcement mechanism external to the players, the equilibrium outcome is defection. Every player is better off if all cooperate. Every player has an individual incentive to defect. The equilibrium is mutual defection. The optimal outcome is cooperation. The market produces the suboptimal outcome.

It is the insight that justified every governance framework described in Chapter 2. The printing press required copyright law because publishers could not voluntarily coordinate to respect each other's rights. Electrical utilities required regulation because companies could not voluntarily coordinate to serve unprofitable rural areas. Nuclear weapons required a treaty because nations could not voluntarily coordinate to limit proliferation. In each case, the coordination problem was solved by an external authority — a government, a regulator, an international institution — that set the rules and enforced compliance.

The AI builders' dilemma requires the same solution. The companies cannot solve it alone. Voluntary self-governance is necessary but insufficient. The solution is governance — rules set by an authority external to the market, binding on all participants, enforced impartially, and designed by a process that includes more than just the companies whose behavior is being governed.

Not because companies are incapable of ethical behavior — some clearly are, and the Anthropic confrontation demonstrates it. But because ethical behavior by individual companies, in a competitive market, is structurally unstable. The room convenes not to replace the ethical actors but to protect them — to ensure that the company that draws a red line is not punished by the market for drawing it, because the red line is the law and the law applies to everyone.

The Anthropic Case Revisited

The Anthropic confrontation deserves a closer examination because it contains, in a single episode, every element of the dilemma and every element of a possible resolution.

The two red lines — no autonomous weapons, no mass surveillance — were not arbitrary. They corresponded precisely to the two applications documented in *The Last Keeper* where the removal of the keeper is most consequential: AI targeting systems that compress human review to a formality (Chapter 9) and surveillance infrastructure that targets the keepers themselves (Chapter 10). Anthropic's red lines were, in the framework of this trilogy, a company saying: we have read the evidence, we understand the cost, and we will not build the tools that remove the keeper from these two loops. The argument for the room. Not a new insight.

The defense establishment's response was also not arbitrary. It reflected a genuine strategic concern: if the companies that build the best AI systems can unilaterally restrict their use by the military, then the military's access to AI capability depends on the ethical commitments of private companies — commitments that can change with a change of leadership, a change of market conditions, or a change of ownership. From the military's perspective, this is an unacceptable dependency. And the dependency is real.

The public response — the million daily sign-ups, the number-one App Store ranking, the surge in users who chose Anthropic specifically because it drew the line — provided a third data point. The market, when given a choice between a company that self-governs and a company that does not, chose the one that self-governs. The finding was not dispositive — consumer preference is not a governance framework — but it was significant. It suggested that the market's response to self-governance may be more complex than the dilemma predicts. The prisoner's dilemma assumes that defection is always rewarded. The public response suggested that, in this case, cooperation was rewarded too.

Whether this holds is an open question. The Anthropic confrontation was a single event, in a specific political context, involving a company with a specific public profile and a specific user base. It does not prove that self-governance is always rewarded by the market. But it demonstrates something the dilemma's formal structure does not capture: the existence of a constituency — millions of users, representing billions of dollars in potential revenue — that values governance. That constituency is a political fact. It is the beginning of the demand side of the governance equation.

> *"The market, given a choice between a company that self-governs and one that does not, chose the one that self-governs. Consumer preference is not a governance framework. But a constituency that values governance is the beginning of the demand side of the equation."*

But this chapter must name a tension in its own argument. The red lines this trilogy celebrates are an exercise of precisely the unilateral corporate power this trilogy warns about. Anthropic's refusal was not a democratic decision. It was a corporate decision, made by a private company's leadership, affecting the military capabilities of a superpower, without legislative authority and without any mechanism that guarantees the red lines will hold beyond the current leadership's tenure. A different board, under different pressures, could draw the lines elsewhere or erase them entirely. lines. It is the argument for the governance framework that would make those red lines a matter of law rather than a matter of corporate courage — which is exactly what the next section proposes.

What the Builders Need

The resolution of the builders' dilemma is not more courageous companies. It is a governance framework that makes courage unnecessary — that converts voluntary ethical commitments into mandatory minimum standards, so that the company that does the right thing is not penalized for doing it. Not an argument against the red

The builders need a floor. Not a ceiling — not a regulatory framework that dictates every aspect of AI development and stifles innovation. A floor: a minimum standard below which no company is permitted to operate. A standard that covers the uses documented in this trilogy as most consequential: autonomous weapons, mass surveillance, deployment in healthcare without validation, deployment in criminal justice without transparency, deployment in critical infrastructure without independent audit. A standard that applies to everyone — so that the company that self-governs above the floor is not competing against companies that operate below it.

The analogy is, again, to the governance of previous technologies. Pharmaceutical companies compete vigorously in the market. They also all comply with FDA approval processes before their products reach patients. The FDA is the floor — the minimum standard below which no company may operate. Above the floor, competition is fierce. Below the floor, competition is prohibited. The floor does not eliminate innovation. It channels innovation in directions that are compatible with the safety of the people the products affect.

The AI industry does not have a floor. It has voluntary commitments from some companies, silence from others, and a competitive environment in which the absence of a floor rewards the companies that spend least on safety and deploy most aggressively. The builders who want to self-govern are asking, in effect, for the room to convene and establish the floor that would protect their ability to compete responsibly.

The Dilemma as Invitation

The builders' dilemma is, paradoxically, the strongest argument for governance — because it comes from the builders themselves. The companies that build frontier AI are not, for the most part, resisting governance. Many of them are requesting it. They are requesting it because they understand the dilemma from the inside: self-governance is insufficient, competitive pressure is relentless, and the alternative to proactive governance is reactive regulation that will be worse for everyone.

The builders are not the only voices that need to be heard. The next chapter asks who else should be in the room — and who is conspicuously absent. The builders have resources, access, and technical expertise. The people who bear the consequences of the builders' decisions — the patients, the defendants, the displaced workers, the surveilled communities, the nations whose sovereignty depends on infrastructure they do not control — have none of these things. And the room, if it convenes, will produce a framework that reflects the interests of whoever is sitting at the table.

The Great Convergence

The Missing Seats

On who is in the room, who is not, and what history teaches about whose voice shapes the rules

"The people most affected by AI are the people least represented in AI governance.
This is not a coincidence. It is a pattern as old as governance itself."

The Missing Seats

On the governance proceedings where the rules are being discussed,the people who are in the room, the people who bear the consequences but have no seat at the table, and whose interests the rules will serve

"The rules always reflect the interests of whoever is in the room. The question is never whether rules will be written. The question is: written by whom? And for whom?"

the lesson of prior governance

The Missing Seats

On who is in the room, who is not, and what history teaches about whose interests the rules will serve

Return, one last time, to Philadelphia.

The fifty-five delegates who gathered in the Pennsylvania State House in the summer of 1787 were, without exception, white men. Most were wealthy. Most were property owners. Many were slaveholders. They were lawyers, merchants, planters, and military officers. They represented twelve states — Rhode Island refused to participate — and within those states, they represented the interests of the propertied class that had selected them.

Women were not in the room. Enslaved people were not in the room — though they were counted, at three-fifths, for the purpose of allocating the political power of the states that enslaved them. Indigenous nations, whose sovereignty the new government would systematically dismantle, were not in the room. Indentured servants, tenant farmers, the landless poor — the vast majority of the population — were not in the room.

The Constitution they produced reflected the interests of the people who wrote it. It protected property rights. It preserved slavery. It restricted the franchise to propertied men. It required two and a half centuries of amendment, litigation, civil war, and social movement to extend its protections toward — not to, but toward — the people who were not in the room when it was written.

framework in history: the rules serve the people who write them. Not exclusively. Not without possibility of amendment. But initially, structurally, and persistently, the framework reflects the interests, the assumptions, and the blind spots of whoever is sitting at the table. The question for AI governance is not only whether the room will convene. It is who will be in the room when it does.

"The Constitution reflected the interests of the people who wrote it. It required two and a half centuries to extend its protections toward the people who were not in the room. The rules always serve whoever is sitting at the table."

Who Is in the Room

The rooms where AI governance is currently being discussed — congressional hearing rooms in Washington, commission chambers in Brussels, summit venues in Bletchley Park and Seoul, boardrooms in San Francisco and Shenzhen — are populated by a specific and remarkably consistent set of actors.

The builders. The CEOs and chief scientists of the companies that develop frontier AI systems testify before Congress, brief regulators, and participate in summits. They bring deep technical expertise and The lesson of every governance

significant institutional resources. They also bring the structural interests of their companies: the desire for regulation that is predictable, that does not disadvantage them relative to competitors, and that does not constrain their ability to develop and deploy their products. Their presence is essential. Their interests are not the only ones at stake.

The lobbyists. In 2023 and 2024, the technology industry spent hundreds of millions of dollars on lobbying related to AI regulation in the United States alone. The lobbying operations employ former congressional staffers, former regulators, and former White House officials who understand the legislative process and have personal relationships with the people who will write the rules. The asymmetry in resources is not subtle: the AI industry's lobbying budget exceeds the total annual budget of most civil society organizations that work on AI governance by orders of magnitude.

The regulators. Government officials tasked with understanding AI and crafting appropriate rules. Many are knowledgeable and committed. Many are also understaffed, under-resourced, and outmatched — attempting to regulate a technology that is evolving faster than their agencies can process, with budgets that are a fraction of the companies' R&D; spending, and with technical expertise that the private sector continually drains through higher salaries. The revolving door between government and industry is well documented and structurally incentivized: the regulator who develops AI expertise becomes more valuable to the companies being regulated than to the agency doing the regulating.

The academics. Researchers from universities and think tanks who provide technical analysis, policy recommendations, and empirical research. Their contribution is invaluable but their influence is constrained by the same resource asymmetry: academic research operates on grant cycles and publication timelines that are measured in years, while the technology and the policy environment change in months.

Who Is Not in the Room

The people who bear the most direct consequences of AI deployment are, with few exceptions, not in the rooms where the rules are being discussed.

The patients. AI diagnostic systems are being deployed in healthcare settings that serve millions of people. The algorithm systematically underserved Black patients by confusing healthcare spending with healthcare need. The patients affected by that algorithm — and by the many similar systems deployed across healthcare — are not represented in AI governance discussions. Patient advocacy organizations exist but are not typically included in the technical and policy forums where AI healthcare regulation is shaped. The clinical perspective is represented by healthcare institutions and professional associations, which are not the same as the patients themselves.

The defendants. AI risk assessment tools influence sentencing, bail, and parole decisions affecting millions of people in the criminal justice system. The defendant Eric Loomis, sentenced to six years of initial confinement based in part on a proprietary risk score he could traced in The Last Keepersystematica

traced in The Last Keeper ?Eric

not examine, challenge, or understand — represents a category of affected person that has no voice in the governance process. Defendants do not testify at AI governance hearings. Incarcerated people do not submit comments to regulatory proceedings. Public defenders, who represent the population most directly affected by algorithmic sentencing, are among the most under-resourced institutions in the justice system.

The displaced workers. The seven hundred Klarna employees from Chapter 6 — and the millions of workers across every white-collar profession who face displacement by AI systems — are not organized, are not represented by institutions with lobbying capacity comparable to the technology companies, and in many cases do not yet know that they are affected. The displacement is distributed across industries, occupations, and geographies in a way that defies traditional labor organizing. There is no union of workers displaced by AI. There is no industry association that represents the economic interests of the people whose judgment is being automated.

The surveilled. The communities monitored by the surveillance infrastructure — the Uyghur population in Xinjiang, the journalists targeted by Pegasus, the activists whose networks are mapped by intelligence agencies, the populations of the Global South whose faces train facial recognition systems they did not consent to — are not in the room. They are, in many cases, in countries that have no effective voice in the international forums where AI governance is discussed.

The Global South. The data that trains frontier AI models is drawn disproportionately from English-language internet content, but the models are deployed globally. The content moderation labor that makes AI systems safe for Western users is performed disproportionately by workers in Kenya, the Philippines, and other low-wage countries. The environmental cost of AI training — the energy consumption, the water usage, the carbon emissions — is borne disproportionately by communities near data centers, which are often located in regions chosen for cheap energy rather than environmental capacity. The nations that contribute the data, the labor, and the environmental cost are not proportionally represented in the governance forums that determine how the technology is built, deployed, and regulated.

> *"The patients. The defendants. The displaced workers. The surveilled. The Global South. The people who bear the most direct consequences of AI deployment are, with few exceptions, not in the rooms where the rules are being discussed."*

The Asymmetry

The asymmetry between those in the room and those absent from it is not an oversight. It is a structural feature of how governance is produced. documented in The Last Keeper

Governance proceedings require resources: the time to monitor regulatory developments, the expertise to understand technical proposals, the staff to draft comments and testimony, the relationships to gain access to decision-makers, and the financial capacity to sustain engagement over the years that governance processes require. The technology companies have these resources in abundance. The affected populations, almost by definition, do not.

The asymmetry is compounded by a feature specific to AI: the affected populations often do not know they are affected. The patient whose healthcare algorithm underestimates their need does not know the algorithm exists. The defendant whose risk score influences their sentence does not know how the score was calculated. The job applicant whose resume was filtered by an AI screening tool does not know the tool was used. The invisibility of algorithmic decision-making — a theme that runs through every chapter of *The Last Keeper* — means that the people most affected by AI are often the people least equipped to advocate for their interests in the governance process, because they do not know their interests are at stake.

It is the governance problem, restated for the current era. The people who bear the consequences of a technology's deployment have always been underrepresented in the governance of that technology. The workers who died in factories were not consulted on factory safety regulations until labor movements forced their inclusion. The communities downwind of nuclear test sites were not consulted on testing policy until decades of illness forced investigation. The pattern is consistent: the affected population is included in governance only after the harm is documented, organized, and politically mobilized — a process that takes years or decades, during which the governance framework is shaped by whoever is already in the room.

The Proxy Problem

The absent parties are not entirely unrepresented. Civil society organizations, academic researchers, investigative journalists, and some government agencies advocate for the interests of the affected populations. These proxies perform essential work. The ProPublica investigation that exposed COMPAS's racial bias. The Obermeyer research that revealed the healthcare algorithm's discrimination. The Citizen Lab investigations that documented Pegasus deployments. The +972 Magazine reporting that described AI targeting systems. Each of these was the work of proxies — researchers, journalists, and advocates who brought the affected populations' experience into the governance conversation.

But proxy representation has structural limitations. The proxies are outspent. The technology industry's lobbying budget dwarfs the combined budgets of every civil society organization working on AI governance. The proxies are outnumbered: a congressional hearing on AI might feature four industry executives and one civil society representative. The proxies are outpaced: the technology evolves faster than the research that documents its effects, which evolves faster than the advocacy that translates research into policy pressure, which evolves faster than the legislative process that converts pressure into law. At every stage of the pipeline from harm to governance, the affected population is behind. Not an AI-specific problem.

And proxy representation carries a deeper problem: it is representation without consent. The civil society organization that advocates for patients affected by AI healthcare bias was not elected by those patients. The researcher who studies algorithmic sentencing was not appointed by the defendants whose liberty is at stake. The proxies are self-selected advocates, often excellent, often effective, but not democratically accountable to the populations they represent. The governance they produce may be better than governance produced without them. It is not the same as governance produced by the people who bear the consequences.

> *"Proxy representation is essential and insufficient. The proxies are outspent, outnumbered, and outpaced. And they represent without consent — self-selected advocates, not elected by the populations whose interests they carry into the room."*

The Precedent for Inclusion

The history of governance expansion is, in every era, the history of missing seats being filled.

The labor movement filled the missing seats in industrial governance. Workers who had no voice in factory conditions organized, struck, and forced the creation of governance mechanisms — unions, labor boards, workplace safety regulations — that included their interests. The process was violent, protracted, and only partially successful. But it produced a governance framework in which the people who bore the consequences of industrialization had, for the first time, an institutional voice.

The civil rights movement filled the missing seats in political governance. Citizens who had been excluded from the franchise — by race, by gender, by economic status — organized, protested, litigated, and forced the extension of governance protections to populations the original framers had excluded. The Fifteenth Amendment. The Nineteenth Amendment. The Voting Rights Act of 1965. Each was the result of sustained political action by people who were not in the room demanding that the room include them.

The environmental movement filled the missing seats in industrial and energy governance. Communities affected by pollution, nuclear testing, and environmental degradation organized, sued, and forced the creation of environmental protection agencies, clean air and water regulations, and environmental impact assessment requirements. The room expanded to include interests that the original occupants had not considered.

In each case, initially serves the people who convene it. The excluded populations organize, mobilize, and force their way in. The governance framework is amended to reflect a broader set of interests. The process takes decades. The harm during those decades falls disproportionately on the excluded.

The AI Challenge

the dynamic recurs: the room

The challenge for AI governance is that the traditional mechanisms for filling missing seats may not operate fast enough.

The labor movement took decades to produce governance frameworks for industrial labor. The civil rights movement took a century to extend the franchise. The environmental movement took decades to produce regulatory frameworks for pollution and environmental protection. In each case, the governance expansion moved at the speed of political organizing — the speed at which affected populations can recognize their shared interest, form organizations, develop political capacity, and exert sustained pressure on governance institutions.

AI deployment is moving at the speed of software releases. The gap between the technology's impact and the affected population's capacity to organize a governance response is wider than at any point in the history of technology governance. The patients affected by biased healthcare algorithms cannot organize a movement at the speed at which new algorithms are being deployed. The workers displaced by AI cannot form a union at the speed at which jobs are being eliminated. The communities subjected to AI surveillance cannot mount a legal challenge at the speed at which the surveillance infrastructure expands.

This means that the initial governance framework for AI — the rules written during the governance window that and closing — will almost certainly be written without adequate representation of the people most affected by the technology. The seats will be missing. The framework will reflect the interests of whoever is present: the builders, the lobbyists, the regulators who understand the technology, and the governments that view AI through the lens of strategic competition.

The question is whether this initial framework can be designed, from the outset, to include mechanisms for its own expansion — built-in pathways for the missing seats to be filled as the affected populations organize. A governance framework that acknowledges its own incompleteness and provides structural mechanisms for amendment is imperfect by design but improvable by structure. A framework that locks in the interests of its initial occupants and provides no pathway for expansion is not governance. It is capture.

"A governance framework that acknowledges its own incompleteness and provides mechanisms for amendment is imperfect by design but improvable by structure. A framework that locks in the interests of its initial occupants is not governance. It is capture."

The Cost of Exclusion

The cost of the missing seats is not abstract. It has been documented, in specific cases, across two books and twenty-two chapters of this trilogy.

When the patient is not in the room, the algorithm confuses cost with need and denies care to the people who need it most. When the defendant is not in the room, the algorithm encodes the biases of a is now open

There is one more missing seat that no governance framework has yet acknowledged, because the seat does not yet exist. Every governance framework in this chapter — every term limit, every succession mechanism, every democratic rotation of power — was designed for holders who eventually die. The biological substrate described in *Who Holds the Jar?*, Chapter 10, is building the technology to change that assumption. CRISPR, senolytics, telomere extension, DNA repair mechanisms borrowed from organisms that exhibit negligible senescence — these are funded, published, and in clinical trials. If the holders of the jar disable mortality, the room's foundational assumption breaks. No framework addresses the seat that never becomes vacant.

There is a category beyond the missing seat: the seat that does not exist. Not a chair left empty at the table, but a population for whom no chair was ever built.

In early January 2026, following a US military operation in Venezuela, the United States effectively severed Cuba's access to imported oil. Cuba had been under a US embargo for over six decades, but oil shipments — primarily from Venezuela and, intermittently, from Russia — had kept the island's infrastructure minimally functional. When that supply line was cut, the consequences were immediate and catastrophic: nationwide blackouts lasting days, the collapse of hospital systems that depend on electrical power for ventilators, refrigeration of medicines, and surgical equipment, and severe food shortages as the cold chain for perishable goods disintegrated.

Eleven million people experienced the downstream consequences of a geopolitical decision in which they had zero voice. Cuba does not attend the World Economic Forum. Cuba is not on the Board of Peace. Cuba is not a party to the semiconductor export controls, the AI governance summits, the quantum computing research consortia, or any of the frameworks where the rules of the computational era are being written. Cuba is not at the table. Cuba is not on the menu. Cuba is in the kitchen — invisible, uncounted, and starving.

> *"There is a category beyond the missing seat: the seat that does not exist.*
> *Cuba is not at the table. Cuba is not on the menu. Cuba is in the kitchen —*
> *invisible, uncounted, and starving."*

framework traced across this trilogy. In New York, you build the Oracle. In Toronto, you negotiate its terms. In Lusaka, you use it without knowing what is behind it. In Havana, you lose your electricity because the Oracle's owners decided your oil supplier was an adversary, and no one in any room, at any altitude, considered your eleven million lives a variable worth including in the calculation.

The missing seats this chapter has d are seats that could, in principle, be filled. The patient, the defendant, the worker, the surveilled community — each has advocates who could represent them, however imperfectly. Cuba represents something worse: a population whose absence from the governance structure is not an oversight. It is a policy. The embargo is the mechanism by which absence is enforced. And the humanitarian cost of that absence — the blackouts, the collapsed hospitals, the hunger — is not a side effect. It is the intended pressure. The invisible population is invisible by design.

catalog

discriminatory justice system and presents the result as objective. When the worker is not in the room, the displacement proceeds without safety nets, retraining frameworks, or transition support. When the surveilled community is not in the room, the surveillance expands without constraint. When the nation whose sovereignty depends on infrastructure it does not control is not in the room, the infrastructure operator makes decisions that determine the outcome of wars.

In January 2026, Canadian Prime Minister Mark Carney stood before the World Economic Forum in Davos and stated the principle that governs every missing seat this chapter has described: "If you're not at the table, you're on the menu." He was speaking about nations — about middle powers forced to choose between hegemons and hyperscalers. But the principle applies at every scale. The patient is on the menu. The defendant is on the menu. The displaced worker is on the menu. The surveilled community is on the menu. The nation whose sovereignty depends on a satellite constellation it does not own is on the menu. The cost of the missing seat is not that your voice goes unheard. It is that the rules are written to serve whoever is present — and you live under rules you had no part in writing.

The missing seats are not an abstraction to be addressed in a future amendment. They are people whose lives are being shaped, right now, by systems they did not consent to, governed by rules they did not write, in a process they cannot access. Every day that the seats remain empty is a day that the rules are written without their input and the consequences fall on them without their voice.

Part IV has now examined all the actors in the governance drama: the builders who face an impossible dilemma (Chapter 9), and the affected populations who are absent from the process (this chapter). The next chapter is where the trilogy converges. Three trajectories — the concentration of power, the removal of the keeper, and the absence of governance — are on a collision course. Chapter 11 maps the moment of convergence and asks what the intersection of all three looks like.

The Great Convergence

The Convergence

On the moment when the trilogy's three trajectories meet, and the question that follows

"The cost documented in Book One, the pattern documented in Book Two, and the governance question posed in Book Three converge in a single moment. This moment."

The Great Convergence

The Convergence

On the moment when the three trajectories of this trilogy meet: power concentrating, the keeper being removed, and governance struggling to keep pace — and the question of what happens at the intersection

"Three curves are converging. Power is concentrating exponentially. The keeper is being removed acceleratingly. And governance is proceeding linearly. The intersection is the present moment."

The Convergence

On the moment when the trilogy's three trajectories meet, and the question of what the intersection produces

This chapter takes its title from the book's title, and it delivers the thesis that the entire trilogy exists to state.

Three trajectories are converging. They have been documented across three books, thirty-eight chapters, and a combined arc that stretches from a shepherd's counting jar in ancient Mesopotamia to an AI system that generates military targets in 2026. Each trajectory has been traced independently. This chapter maps their intersection.

The first trajectory is the concentration of power. Documented across fourteen chapters of *Who Holds the Jar?*, this trajectory shows that computational power has concentrated in fewer hands with every era — from the temple administrators who controlled the counting tokens, through the governments that controlled the mainframes, the companies that controlled the operating systems and processors, the platforms that controlled the internet, to the seven organizations that can train frontier AI models, the two companies that manufacture the chips, and the one company that controls sixty percent of active satellites. The pattern is distribute-then-recapture: each era distributes capability and then reconcentrates control. The concentration is accelerating. The current era's reconcentration is happening simultaneously with its democratization, leaving no gap for governance to intervene.

The second trajectory is the removal of the keeper. Documented across twelve chapters of *The Last Keeper*, this trajectory shows that the human in the loop — the person who exercises judgment when the system encounters conditions it was not designed for — is being removed, degraded, overridden, deauthorized, outpaced, and overwhelmed every domain where automated systems make consequential decisions. The removal is justified each time by , cheaper, more consistent. The reasoning is correct on average. The cost is paid at the margin — in lives, liberty, and justice.

The third trajectory is the absence of governance. Documented across the ten chapters of this book that precede this one, this trajectory shows that the governance frameworks that could arrest the first two trajectories — that could constrain the concentration and preserve the keeper — are fragmentary, inadequate, outpaced by the technology, captured by the builders, and missing the voices of the people most affected. The room has not convened. The rules have not been written. And the window is closing.

"Power is concentrating exponentially. The keeper is being removed acceleratingly. Governance is proceeding linearly. The three curves are converging. The intersection is now."

familiar reasoning: the machine is faster

The First Convergence: Capability and Governance

AI capability and AI governance are on a collision course.

Capability is advancing exponentially. The performance of frontier AI models on standard benchmarks has improved at rates that consistently exceed predictions. Tasks that AI could not perform two years ago — writing competent code, passing professional licensing examinations, generating photorealistic images, conducting multi-step reasoning — are now routine. The capability frontier moves forward in months. The improvements are driven by increases in compute, data, and training methodology, all of which are following trajectories that show no sign of approaching a ceiling.

Governance is proceeding linearly — through legislative processes, regulatory consultations, international negotiations, and institutional adaptations designed for a world that moved at a different speed. The EU AI Act took approximately three years from proposal to enactment. The US has not passed comprehensive AI legislation. International coordination on AI governance is in its earliest stages. Each of these processes operates at the tempo of human institutions: hearings, comment periods, committee markups, floor votes, diplomatic negotiations, treaty ratifications. The tempo is measured in years.

An exponential curve and a linear curve will always intersect. When they do, the exponential curve pulls away — the gap between capability and governance widens, and the governance becomes progressively less relevant to the technology it is attempting to govern. The rules written for today's AI will not address next year's AI. The framework designed for current capabilities will be obsolete by the time it is enacted. The convergence of capability and governance is not a future event. It is the present condition.

The Second Convergence: The Lego and the Lumber

Within the broader AI governance challenge, there is a specific convergence that affects every organization that builds or uses software — which is to say, every organization on earth.

The distinction introduced in Chapter 4 — between model-driven development (Lego bricks with built-in safety) and code-generated development (raw lumber without constraints) — is not an abstract architectural preference. It is the front line of the governance battle, and it is being fought inside every enterprise, every government agency, and every institution that depends on software.

On one side: platforms that enforce structure. Enterprise software built on model-driven architectures that constrain what the builder can create to configurations the platform has validated. The safety is in the system, not in the individual. The platform catches the error the builder does not see. The governance is embedded in the tool.

On the other side: AI tools that generate raw code from natural-language descriptions, with no platform enforcing structure, no automated auditing, no compliance verification, and no mechanism to catch the vulnerabilities that the builder cannot detect and the tool does not flag. The capability is striking. The governance is absent.

The convergence is this: both approaches are advancing simultaneously, and the organizations that must choose between them — or manage some combination of both — are making decisions that will determine the security, reliability, and auditability of the software that runs healthcare systems, financial markets, government services, and critical infrastructure for the next decade. The choice between Lego and lumber is not a product decision. It is a governance decision. And it is being made, in most organizations, without a governance framework to guide it.

> *"The choice between Lego and lumber is not a product decision. It is a governance decision. It is being made in every organization on earth, without a governance framework to guide it."*

This convergence is a microcosm of the larger one. At the enterprise level, the question is: will we build software with systems that enforce safety or systems that leave safety to the builder? At the civilizational level, the question is the same: will we develop AI within frameworks that enforce accountability or in a vacuum that leaves accountability to the builder? The answers at both scales will be determined by whoever writes the rules — or by the absence of rules, which produces its own answer.

The Third Convergence: The Trilogy

The three books of *The Cost of the Machine* converge in this chapter, and the convergence is not metaphorical. It is visible in a single case study.

In September 2022 and February 2026, one company — SpaceX — demonstrated all three trajectories simultaneously.

Power concentration (*Who Holds the Jar?*): Starlink controls over sixty percent of active satellites in low Earth orbit. The capability to provide orbital internet connectivity is concentrated in one company to a degree that has no parallel in the history of communications infrastructure. The distribute-then-recapture pattern: anyone can buy a terminal, one company owns the constellation.

The removal of the keeper (*The Last Keeper*): the decision to grant or deny connectivity — a decision that determines whether military operations succeed, whether hospitals stay online, whether a nation can defend itself — is made by one person, without independent review, without democratic mandate, without the governance framework that would place a keeper between the decision and its consequence. The keeper in this case is not removed from the loop. The keeper *is* the loop — one individual holding a switch that the governance system has not learned to govern.

The absence of governance (*The Great Convergence*): no treaty governs private orbital communications monopolies during armed conflict. No international body has jurisdiction. No governance framework requires the operator to maintain service, prohibits the operator from denying service, or provides a mechanism for review after either decision. The room has not convened for this scenario because the scenario did not exist until one company built a constellation that two sides of a war both

came to depend on.

One company. One constellation. One person's decisions. Three trajectories — power, cost, and governance — intersecting in a case study that is not hypothetical, not projected, not modeled. It has already happened. The convergence is not an event to be anticipated. It is a condition to be recognized.

> *"One company. One constellation. One person's decisions. Three trajectories*
> *— power, cost, and governance — intersecting in a case study that has*
> *already happened. The convergence is not an event to be anticipated. It is a*
> *condition to be recognized."*

A second convergence, at the level of nations, became visible within days of this book's writing. When Canadian Prime Minister Mark Carney named the rupture in the rules-based order at Davos in January 2026, the response demonstrated all three trajectories simultaneously. Power concentration: the hegemon used trade threats, tariff escalation, and the deliberate encouragement of internal separatism to punish a sovereign nation for naming reality. The removal of the keeper: the leader who exercised judgment — who said the thing the system needed said — was threatened with consequences designed to deter future truth-telling, not just by Canada but by every middle power watching. The absence of governance: no international framework protects a nation from economic retaliation for speech at a diplomatic forum. The rules that would have prevented the punishment do not exist — because the rules-based order whose protection they would have invoked is the very system Carney was declaring ruptured.

The Starlink case is a convergence of infrastructure. The Carney case is a convergence of governance. Together they demonstrate that the three trajectories are not converging in a single domain. They are converging everywhere — in technology, in diplomacy, in the relationship between nations and the infrastructure they depend on. The convergence is systemic.

The Five Phases Applied to AI

The five-phase cycle traced across four technologies in Chapters 2 and 3 can now be applied to artificial intelligence with the specificity this chapter's position in the book demands.

Phase 1 — Innovation (2012–2020): The deep learning breakthroughs. AlexNet's image recognition in 2012. DeepMind's AlphaGo defeating Lee Sedol in 2016. GPT-2 demonstrating coherent text generation in 2019. The capability existed. It was expensive, concentrated in a handful of research labs, and inaccessible to the general population. Complete.

Phase 2 — Democratization (2022–present): ChatGPT, November 30, 2022. One hundred million users in two months. Open-source models by 2023. AI code generation tools by 2024. The Shopify mandate by 2025. The capability distributed to hundreds of millions of people. Accelerating.

Phase 3 — Chaos (2024–present): Documented across Part II of this book. Ungoverned software creation at scale (Chapter 4). The epistemic crisis of AI-generated disinformation (Chapter 5). White-collar labor displacement at unprecedented speed (Chapter 6). The harms documented across *The Last Keeper* — biased algorithms, algorithmic amplification, AI targeting, mass surveillance — intensifying. Beginning.

Phase 4 — Consolidation (2023–present): Documented across Part III of this book. Seven frontier model builders. Two chip companies. Three cloud providers. One orbital communications monopoly. The vertical integration of the AI stack. The geopolitical competition between two superpowers. The distribute-then-recapture pattern operating in real time. Already visible.

Phase 5 — Governance: Fragments. The EU AI Act. Executive orders. National strategies. The Bletchley Declaration. Corporate self-governance that is structurally unstable. No comprehensive international framework. No Philadelphia. No Bretton Woods. No NPT. Incomplete.

The critical observation: Phases 3 and 4 are happening simultaneously. cycle, democratization produced chaos, and the chaos produced consolidation, in sequence. AI is producing chaos and consolidation at the same time. The builders who are consolidating control of the infrastructure are also the builders whose tools are producing the chaos of ungoverned deployment. The entities that would need to be governed are the same entities that are consolidating the power that makes governance difficult. This simultaneity has no precedent in the historical cases . It means the governance window is not just short. It is closing from both directions — compressed by the chaos that demands governance and by the consolidation that resists it.

> *"Phases 3 and 4 are happening simultaneously. Chaos and consolidation at the same time. The window is closing from both directions — compressed by the chaos that demands governance and by the consolidation that resists it."*

The Indicators

How would we know if the governance window is closing? What are the indicators that would signal the transition from a window that is open — where governance is still possible — to a window that has shut — where the consolidation is irreversible and the rules will be written by the consolidators?

The printing press provides the template. The window closed when state censorship regimes became the primary mechanism of information control — when the consolidators (governments) wrote the rules (censorship) and the rules served the consolidators' interests (suppression of dissent). For AI, the indicators of a closing window would include:

The achievement of self-sufficiency in AI compute by a major power, eliminating the chokepoints that currently provide leverage. The establishment of de facto technical standards by dominant AI companies that become impossible for regulators to override. The embedding of AI systems so deeply in critical infrastructure that removing or modifying them becomes practically impossible. The normalization of the historical record shows In prior technology cycles

AI-driven decision-making in domains like healthcare, criminal justice, and military operations to the point where the demand for human oversight is treated as impractical rather than essential. The political capture of regulatory institutions by the companies they are tasked with governing. And the failure of international coordination — the moment when it becomes clear that national approaches have fragmented to the point where a coherent global framework is no longer achievable.

Not all of these indicators have been triggered. Some are well advanced. The embedding of AI in critical infrastructure is accelerating. The normalization of AI-driven decision-making is proceeding in healthcare, criminal justice, and military operations. China's drive toward compute self-sufficiency is underway. The fragmentation of governance approaches is visible. The window is not shut. But it is closing, and several of the indicators that would signal its closure are trending in the wrong direction.

The Present Moment

We are, as of this writing in March 2026, in the governance window for artificial intelligence. The window is open. The question this chapter exists to answer is: what would it take to keep it open long enough for the room to convene?

The historical precedents suggest three requirements.

First: a shared recognition of urgency. The NPT was made possible by the Cuban Missile Crisis — thirteen days that made the risk undeniable. The Geneva Conventions were made possible by two world wars. The printing press's governance arrived after 130 years of religious warfare. every case, governance was catalyzed by a crisis that made the cost of inaction intolerable. The question is whether AI governance can be catalyzed by the evidence already available — the costs documented in *The Last Keeper*, the concentration documented in *Who Holds the Jar?*, the chaos documented in this book — or whether it will require a crisis of a magnitude not yet seen. The hope of this trilogy is that the evidence is sufficient. The precedent is not encouraging.

Second: a coalition willing to act. The NPT required the cooperation of the two superpowers — adversaries who agreed on one thing: uncontrolled proliferation was a shared threat. AI governance requires a coalition that includes the major AI-developing nations, the companies that build the technology, the civil society organizations that represent the affected populations, and the international institutions that can provide coordination and enforcement. No such coalition currently exists. Elements of it are visible: the Bletchley Declaration, the bilateral discussions between the US and China on AI safety, the EU's regulatory leadership. But a declaration is not a coalition. Discussions are not agreements. Regulation in one jurisdiction is not governance of a global technology.

Third: a framework designed for speed. for a technology that had stabilized. AI has not stabilized and may not stabilize. The governance framework must be adaptive — designed not for a specific capability but for a class of capabilities that evolves while the rules are being applied. This has no precedent. The closest analogue is financial regulation, which Prior governance frameworks were designed

attempts to govern a system — global finance — that evolves continuously. The track record of financial regulation is mixed: it has prevented some crises and failed to prevent others. But it demonstrates that adaptive governance is at least conceptually possible, even for systems that outpace the regulators.

The Moment

on.

Power is concentrating. The keeper is being removed. Governance is absent. The three trajectories are not parallel — they are reinforcing. The concentration of power makes governance harder, because the entities that would be governed have the resources to resist it. The removal of the keeper makes governance more urgent, because the cost of ungoverned systems is measured in lives. And the absence of governance accelerates both the concentration and the removal, because neither is constrained.

The reinforcing dynamic is the reason the convergence is dangerous. Each trajectory amplifies the others. Concentrated power removes keepers faster (because the entities with power face less accountability). Removed keepers make concentration easier (because the humans who might challenge the concentration are no longer in the loop). Absent governance allows both to proceed unchecked. The triangle tightens.

The final chapter makes the closing argument. It returns to the room. It names the choice. And it asks whether the evidence documented across three books and thirty-eight chapters — five thousand years of computational history, a century of catastrophic failures, and the governance patterns of every transformative technology — is sufficient to convene the room before the window closes.

The moment the trilogy converges

The reinforcing dynamic has a human measure. In 1983, Stanislav Petrov saved the world because he had twenty minutes and the authority to override a machine. The Soviet early-warning system reported five American nuclear missiles inbound. Petrov judged the data inconsistent with a genuine first strike and declared it a false alarm. He was right — sunlight on high-altitude clouds had fooled the sensors. The keeper worked because three conditions held: he had time, he had context, and he had authority. The Lavender targeting system deployed forty years later gives its operator twenty seconds and a rubber stamp. The governance question is not whether human judgment matters — Petrov proved it does. The question is whether any framework will require it.

The Great Convergence

The Choice

On September 17, 1787, the final day of the Constitutional Convention, and the question this trilogy leaves with the reader

"The window is open. It will not stay open. The question is whether the room will convene before it shuts."

The Choice

The choice is not between perfect governance and imperfect governance. It is between imperfect governance and none.

The history documented across this trilogy provides no example of a transformative technology that was governed perfectly. The Constitution sanctioned slavery. The Geneva Conventions are violated in every war. The NPT has not eliminated nuclear weapons. The EU AI Act is a framework for one continent in a world where the technology is planetary. Every governance framework in history has been imperfect, incomplete, shaped by the interests of whoever was in the room, and amended over time as the excluded populations forced their way to the table.

The history also provides no example of a transformative technology that was better off ungoverned. The printing press, ungoverned for 130 years, produced the Wars of Religion. Electricity, ungoverned for decades, produced monopoly empires that collapsed on their customers. The internet, never adequately governed, produced surveillance capitalism, platform monopolies, and a recommendation algorithm that played a determining role in genocide. The absence of governance was not an absence. It was a presence — the presence of the consolidators' rules, written in terms of service, in algorithmic design, in corporate policy, without democratic mandate and without accountability to the people who lived under them.

The choice for artificial intelligence is the same choice that every transformative technology has eventually presented. Not whether to govern — every powerful technology has been governed, sooner or later, well or badly. But when. And by whom. And at what cost in the interim.

The Window

The governance window for AI is open.

The printing press took 350 years to govern. The chaos phase killed millions. Electricity took 53 years. The chaos phase produced monopoly abuse and financial collapse. Nuclear energy took 23 years. The chaos phase nearly destroyed civilization. The internet was never adequately governed. The chaos phase is ongoing.

The Great Convergence

Chapter 12 of 12

The Choice

The closing argument. The return to the room. And the question the trilogy ends on.

"I confess that there are several parts of this constitution which I do not at present approve, but I am not sure I shall never approve them."— Benjamin Franklin, September 17, 1787

The Choice

On September 17, 1787, the final day of the Constitutional Convention, Benjamin Franklin rose to address the delegates. He was eighty-one years old, the oldest person in the room, too frail to deliver his remarks himself. He handed his written speech to James Wilson of Pennsylvania, who read it aloud.

Franklin began by acknowledging that the document before them was imperfect. He confessed that there were parts of it he did not approve. He noted that he was not sure he would never approve them — a gentle admission that his own judgment was fallible and that time might reveal wisdom in provisions he currently doubted. He observed that every delegate who came to the convention had brought their prejudices, their passions, their errors of opinion, and their local interests, and that the document reflected all of these as much as it reflected the delegates' collective wisdom.

And then he urged them to sign it.

Not because it was perfect. Not because it represented the best governance framework that could theoretically be designed. But because it was the best framework that this group of imperfect people, with their imperfect knowledge and their imperfect motives, could produce in this room, at this moment, under these constraints. And because the alternative — no framework, no governance, no room — was worse.

Franklin's speech was not a celebration. It was a clear-eyed assessment of what governance actually looks like when human beings do it: compromised, incomplete, stained by the interests of whoever happens to be present, and incomparably better than nothing.

> *"The document was imperfect. Franklin urged them to sign it anyway — not because it was the best framework that could theoretically be designed, but because it was the best this group of imperfect people could produce at this moment. And because nothing was worse."*

The Trilogy's Argument

This is the final chapter of the final book of *The Cost of the Machine*. The trilogy's argument can now be stated in full.

The Last Keeper documented the cost of removing the human from the loop. It traced the pattern from a lighthouse on a rock off Pembrokeshire through radiation therapy rooms, cockpits, courtrooms, clinics, social media platforms, military targeting systems, surveillance infrastructure, and software too complex for any human to verify. — the human who exercised judgment when the system encountered conditions it was not designed for — was removed, degraded, overridden, deauthorized, outpaced, or overwhelmed. In every case, the removal was justified by is better on average. In every case, the cost was paid at the margin — in lives, in liberty, in justice. The keeper's question: will there be a human watching the light when the mechanism fails? The keeper

familiar reasoning: the machine is better on ave

Who Holds the Jar? documented the pattern that produces the cost. It traced the concentration of computational power across five millennia — from the shepherd's counting jar through the abacus, the clockwork calculator, the mechanical engine, the electronic switch, the transistor, the microprocessor, the personal computer, the internet, and the neural network. every era, same pattern repeated: capability distributed, then control reconcentrated. The distribute-then-recapture cycle produced a ratchet of power that tightened with each turn. The jar's question: who will hold the computational power of the next era — and will it be possible to take it back?

The Great Convergence documented the only force capable of arresting both patterns: governance. It traced the five-phase cycle that every transformative technology has followed — Innovation, Democratization, Chaos, Consolidation, Governance — across the printing press, electricity, nuclear energy, and the internet. It showed that the governance window is shrinking with each era, that the chaos phase of AI has begun, that the consolidation is already advanced, and that the room has not yet convened. The convergence's question: who will write the rules — and what happens when nobody does?

Three books. Three questions. One argument: we build machines that accumulate power, we remove the humans who could check those machines, and nobody writes the rules to govern what we have built. The cost is measured in lives, liberty, sovereignty, and the capacity of societies to govern themselves. The pattern has repeated in every era. The question is whether it repeats in this one.

> *"We build machines that accumulate power. We remove the humans who*
> *could check those machines. Nobody writes the rules. The cost is measured in*
> *lives. The pattern has repeated in every era. The question is whether it repeats*
> *in this one."*

The False Choices

The debate about AI governance, as conducted in the public sphere, is dominated by false choices that this trilogy's evidence should put to rest.

The false choice between innovation and governance. The printing press was governed and continued to innovate. Electricity was governed and continued to power civilization. Nuclear energy was governed and continues to provide clean power and scientific capability. Aviation is governed by one of the most comprehensive regulatory frameworks in existence and is the safest form of transportation in human history. In no case did governance eliminate innovation. innovation in directions compatible with human welfare. The argument that AI governance will stifle innovation is not supported by a single historical example. The argument that ungoverned AI will produce catastrophe is supported by every example

The false choice between governance and speed. The argument that governance is too slow for AI is an argument against bad governance, not against governance itself. The NPT was negotiated in under Governance channeled

documented across thirty-eight chapters. The false

three years once the political will existed. Aviation safety regulations are updated continuously as technology evolves. Financial regulation adapts to new instruments and new risks on an ongoing basis. The governance framework for AI does not need to be a static document that takes a decade to write. It needs to be an adaptive framework — a set of principles, institutions, and enforcement mechanisms that evolve with the technology. The challenge is real. The challenge is not a reason to abandon the attempt.

The false choice between national sovereignty and international coordination. AI governance does not require nations to cede sovereignty. It requires nations to coordinate, as they have for nuclear weapons, for aviation safety, for financial stability, for climate change. Coordination is not surrender. It is the recognition that a planetary technology requires a planetary response, and that national governance of a global technology is an oxymoron.

The false choice between perfect governance and no governance. the false choices, because it disguises inaction as prudence. The argument goes: we do not yet know enough about AI to govern it well, so we should wait until we know more. The printing press was not well understood when the Statute of Anne was passed. Nuclear physics was not fully understood when the NPT was signed. Governance is never written with complete knowledge. It is written with the best knowledge available, and it is amended as knowledge improves. The alternative to imperfect governance is not better governance later. The alternative is the consolidation of power without constraint, the removal of the keeper without check, and the chaos phase without limit — the precise conditions documented across thirty-eight chapters.

The false choice between naming reality and surviving it. In January 2026, Mark Carney stood before the most powerful economic forum in the world and said the thing that the room had been performing around for years: the rules-based order is ruptured, and the nations that pretend otherwise are accepting subordination while calling it sovereignty. The cost was immediate and visible. Revocation from the Board of Peace. Threats of 100% tariffs. The deliberate encouragement of provincial separatism by a foreign government. The Treasury Secretary of the hegemon describing a Canadian province as "a natural partner for the US" while meeting with groups working to fragment the nation. And the attempted rewriting of what was said — the claim that Carney was "aggressively walking back" remarks he publicly reaffirmed. The cost of naming reality was real. But what Carney gained was something the cost could not erase: a room. A standing ovation from world leaders. The President of Finland calling it one of the best speeches heard at Davos. The Secretary General of NATO saying "Canada is back." And most critically, a framework that other middle powers could organize around — the TPP-EU trade bridge, the G7 critical minerals buyer's clubs, the AI governance cooperation among like-minded democracies. Carney did not win the argument with the hegemon. He did something more important: he convened the room that the hegemon's dominance had prevented. Imperfect. Incomplete. Exactly like Philadelphia. And incomparably better than the alternative, which was silence. The false choice says: name reality or survive. The evidence says: naming reality is how the room convenes — because someone has to speak first, and the constituency that values truth does not know it exists until someone gives it a voice.

documented across

Philadelphia is the right model not because it produced perfect governance but because it produced *amendable* governance. The Constitution that excluded most of humanity — no women, no enslaved people, no unpropertied men — also contained the mechanism for its own correction. The amendments came. They came slowly, at tremendous cost, and they are still incomplete. But they came — because the document was written to allow revision. The AI governance that this trilogy calls for must do the same: write the rules now, with the people available, and build in the amendment process that allows the missing seats to reshape the framework once they arrive.

What Governance Requires

The evidence documented across this trilogy points to a governance framework with five elements.

A floor, not a ceiling. Minimum standards below which no AI developer may operate, covering the most consequential domains: autonomous weapons, mass surveillance, deployment in healthcare without validation, deployment in criminal justice without transparency, deployment in critical infrastructure without independent audit. Above the floor, competition is unrestricted. Below the floor, competition is prohibited. The floor protects the companies that self-govern from being undercut by those that do not.

Keepers in the loop. Mandatory human oversight in domains where the cost of failure is measured in lives — the domains documented across *The Last Keeper*. Not human oversight as a checkbox that can be compressed to twenty seconds per target. Meaningful human oversight, with the time, the authority, the information, and the institutional support to exercise independent judgment. The keeper must be present. The keeper must be empowered. The keeper must be practiced. The governance must require all three.

Transparency at the point of consequence. The people affected by algorithmic decisions must be able to know that an algorithm was used, how it was used, and on what basis. The defendant must be able to challenge the risk score. The patient must be able to know that an algorithm influenced their treatment plan. The job applicant must be able to know that an AI system screened their resume. Transparency does not require disclosing proprietary algorithms. It requires disclosing that an algorithm was used, what its inputs were, and how the affected person can contest the outcome.

International coordination. A framework that matches the technology's scope. The AI equivalent of the IAEA — an international institution with the mandate to monitor frontier AI development, the technical capacity to evaluate risks, and the authority to coordinate national governance efforts. The institution does not need to be perfect. The IAEA is not perfect. It needs to exist.

Seats for the affected. Structural mechanisms that ensure the people who bear the consequences of AI deployment have a voice in its governance. Patient representatives in healthcare AI oversight. Defendant advocates in criminal justice AI policy. Worker representatives in AI labor transition planning. Civil society organizations from the Global South in international AI governance forums. The seats will be imperfect. The representation will be incomplete. But the alternative — a framework written exclusively by the builders and the governments that fund them — will reflect exclusively their interests, as every

The window for AI is measured in years, not decades. The capability is advancing monthly. The consolidation is already advanced. The chaos is beginning. The room has not yet convened in a form adequate to the challenge.

The evidence is sufficient. Three books have documented the cost: lives lost to systems whose keepers were removed, power concentrated , and governance frameworks that arrived too late for the printing press, too late for electricity, too late for the internet, and nearly too late for nuclear weapons. The evidence does not say that AI governance will succeed. It says that the attempt must be made. It says that the room must convene. It says that imperfect rules, written now, by imperfect people, in an imperfect process, are incomparably better than the alternative.

And the alternative is not an absence. It is a choice — made by default, by the people who benefit from the absence, at the expense of everyone else.

> *"The evidence does not say that AI governance will succeed. It says that the attempt must be made. It says that imperfect rules, written now, by imperfect people, in an imperfect process, are incomparably better than the alternative."*

The Jar. The Light. The Room.

Eight thousand years ago, a shepherd on the plains of Mesopotamia pressed a clay token into a jar. The token represented a sheep. The jar was the first ledger. The ledger was the first computer. And the question — who holds the jar? — has echoed across every era since, growing louder as the jar grew more powerful and the stakes of holding it grew higher.

Two hundred years ago, a keeper on a rock in the Atlantic kept a light burning through storms, beside the body of his dead companion, because the light was not for him. The light was for the ships. And the question — will there be a keeper watching the light when the mechanism fails? — has echoed through every automated system that removed the human who could catch the failure the machine could not see.

Two hundred and thirty-seven years ago, fifty-five men gathered in a room in Philadelphia, nailed the windows shut, argued for four months, and produced an imperfect document that has governed the most powerful nation on earth ever since. And the question — who is in the room, and who writes the rules? — has echoed through every governance moment since, from Bretton Woods to Geneva to the NPT to the room that never convened for the internet.

Three images. Three questions. One argument.

The jar is filling with a power no previous era has contained. The light is burning, but the keeper is being told the light does not need watching. And the room — the room where the rules for this power and this light could be written — has not yet convened.

The window is open. The jar is filling. The keeper is watching. The room is waiting. unchecked

Who will be in the room? What rules will they write? And will the rules arrive before the window closes?

> *"The choice is not between perfect governance and imperfect governance. It is between imperfect governance and none. And if the history documented teaches anything, it is that none is not an absence. It is a choice — made by default, by the people who benefit from the absence, at the expense of everyone else."*

the documented evidence teaches

The Cost of the Machine

The Last Keeper

Who Holds the Jar? *The Machine Evolves: A Natural History of Computation — and a Warning*

The Great Convergence

Three books. One argument. The jar is filling. The keeper is watching. The room is waiting.

Pumulo Sikaneta

References & Source Attribution

Additional References

About the Author

Pumulo Sikaneta works in technology strategy, where he has spent years studying how organizations adopt, govern, and are transformed by the systems they build. His professional work in partner enablement and solutions architecture has given him a close-up view of the gap between what technology promises and what it delivers — and of the humans who stand in that gap.

The Great Convergence is the third and final book in *The Cost of the Machine*, a trilogy that began with *The Last Keeper: Who Watches the Light? — On the Human Cost of Trusting Machines* and continued with *Who Holds the Jar? — The Machine Evolves: A Natural History of Computation and a Warning*. Together, the three books trace a single argument from the human cost of removing people from the loop through the five-thousand-year pattern of computational power concentration to the governance battle that will determine whether the pattern repeats or breaks.

The inspiration for this trilogy came, in part, from conversations with his brother, Dr. Tabo Sikaneta, a nephrologist at the Scarborough Health Network in Toronto and founding member of the African Caribbean Kidney Association. Dr. Sikaneta's two decades of clinical work documenting the disproportionate burden of kidney disease in immigrant and non-White communities — and his 2025 *BMJ Open* study demonstrating that these disparities persist even within universal healthcare — provided the human ground truth that animates this trilogy's argument: that the data our systems learn from carries the inequities of the world that produced it, and that no algorithm can be more just than the society whose data it consumes.

The author was born in Zambia, raised in Canada, and lives in the United States — three countries with three different relationships to the power structures this trilogy examines. He believes that the most important skill in any field is the willingness to be corrected.

Afterword

What didn't make it. What to watch.
And what's at stake if nobody does.

What Didn't Make It

Three books. Thirty-eight chapters. And I am haunted by what I left out.

The US military operation in Venezuela in January 2026 — the raid that captured Nicolás Maduro, killed eighty-three people including forty-seven Venezuelan soldiers, and reportedly used an AI system through a defense contractor's classified platform — deserved its own chapter. It is the most concrete case yet of a frontier AI model deployed in a kinetic military operation, and the fallout between the AI company and the Department of War over how the technology was used is a live demonstration of every tension this trilogy documented: the keeper's red lines tested in real combat, the infrastructure gatekeeper threatened with blacklisting for asking questions, and the governance vacuum that leaves no framework for adjudicating any of it.

Cuba's humanitarian crisis deserved its own section — a nation under six decades of embargo whose population is starving while the geopolitical powers that imposed the embargo and the powers that claim to oppose it use the island's suffering as a bargaining chip. Cuba is a case study in what happens when a population has no seat at any table: the consequences are borne entirely by people who have no voice in the decisions that produce those consequences. The missing seat, rendered in human hunger.

China's expansion of AI-powered surveillance beyond Xinjiang into language-specific monitoring systems deserved deeper treatment than   10 of *The Last Keeper* could provide. Chinese companies are now developing large language models specifically for Uyghur, Tibetan, and Mongolian — not to serve those communities but to monitor and control their communications with a precision that previous surveillance infrastructure could not achieve. The "happy Uyghurs" deepfake campaign — fabricated videos depicting contented lives in Xinjiang, distributed at scale to counter documented evidence of repression — is the liar's dividend from Chapter 5 of *The Great Convergence* operating as state policy.

Russia's information operations deserved a chapter of their own. The coordinated use of AI-generated deepfakes in election interference — across thirty-eight countries, affecting 3.8 billion people since 2021 — and the deliberate contamination of Western AI training data by flooding search results with state-sponsored narratives, represent a new category of epistemic attack: not targeting human readers directly but poisoning the AI systems that humans increasingly rely on for information. The machine itself is being taught to lie.

The hyperscalers' labor practices deserved more than a single chapter. The content moderation workers in Kenya and the Philippines, documented briefly in *The Great Convergence*, are the human infrastructure of the AI economy — underpaid, psychologically traumatized, and invisible to the users whose experience their suffering maintains. Meta's open-source releases, which this trilogy examined as a governance question, are simultaneously a labor question: the models are trained on data labeled by workers earning less than two dollars an hour, and the open-source release ensures that no downstream governance can protect those workers or the communities whose data they processed.

Climate AI, autonomous vehicles, AI in education, the environmental cost of training runs, AI-generated child exploitation material, the psychological effects of AI companionship systems — each is a chapter this trilogy did not write. The evidence is accumulating faster than any author can document it. The cost is being paid in domains this trilogy did not reach.

"The evidence is accumulating faster than any author can document it. The cost is being paid in domains this trilogy did not reach."

What to Watch

The three books of this trilogy rest on different evidentiary foundations, and the difference matters for how the reader should hold them.

The Last Keeper is built on historical evidence. The Therac-25 killed patients. Air France 447 crashed. The 737 MAX killed 346 people. COMPAS encoded racial bias. Facebook's recommendation engine amplified genocide. The Ariane 5 exploded. These events happened. They are documented. The keeper's patterns — the Reliability Trap, the Single Point of Trust, the Invisible Bias, the Speed-Judgment Tradeoff, the Accountability Gap — are derived from evidence that has already been tested by the world. If the first book is wrong, it is wrong about interpretation, not about fact.

The Great Convergence is built on evidence that is in flight. The governance window is a framework applied to a process that is still unfolding. The five-phase cycle has completed for the printing press, electricity, and nuclear energy. For AI, we are in Phase 3 and Phase 4 simultaneously, and Phase 5 has not arrived. The Starlink case studies, the Carney aftermath, the Anthropic confrontation, the semiconductor export controls — these are events that happened, but their consequences are still developing. If the third book is wrong, it may be wrong about the trajectory: the window may be wider than I believe, or the consolidation may be less entrenched, or the governance may arrive faster than the historical precedents suggest.

Two numbers arrived as this afterword was being written that make the omissions concrete.

On the environmental cost: a February 2026 report commissioned by Beyond Fossil Fuels found that 74 percent of Big Tech's claims regarding AI's climate benefits are unproven — characterising them as a "hoax" designed to greenwash the massive fossil fuel demands of generative AI data centres. The companies that build frontier models are simultaneously claiming that AI will solve climate change and consuming electricity at a rate that accelerates it. The environmental cost of the compute infrastructure is not a side effect of the AI era. It is a structural feature — and the gap between the industry's climate rhetoric and its energy consumption is itself a governance failure that this trilogy did not have room to examine.

On AI-generated child exploitation material: the Internet Watch Foundation documented, in early 2026, a 26,362 percent increase in photo-realistic AI-generated videos of child sexual abuse over the previous year. That number is not a typo. Twenty-six thousand percent. This is the chaos phase of the five-part cycle described in The Great Convergence, Chapter 2, operating in its most obscene domain — a technology democratised without governance, producing harms at a scale and speed that existing legal frameworks cannot contain. The children in these generated images do not exist. The market for them does. The harm to real children — through normalisation, through the training of predatory   , through the overwhelming of detection systems designed for a pre-AI volume of material — is not hypothetical. It is measured, documented, and accelerating.

Who Holds the Jar? is built on something different from the other two. It is built on a pattern observed across five thousand years — the distribute-then-recapture cycle — and an extrapolation of that pattern into substrates that do not yet exist at scale: quantum computing and biological computation. The historical pattern is documented. The extrapolation is a warning, not a certainty. The book argues that quantum and biological computing may produce a concentration of computational power so profound that it cannot be reversed — that a small number of entities may gain the ability to compute things the rest of the world cannot even verify. This has not happened yet. It is not what most people are concerned about today. The public conversation is about chatbots, deepfakes, and job displacement — the chaos phase documented in *The Great Convergence*. The substrate question is further away, less visible, and consequently less discussed.

It is also the question that changes everything.

> *"The public conversation is about chatbots and deepfakes — the chaos phase. The substrate question is further away, less visible, and consequently less discussed. It is also the question that changes everything."*

If the distribute-then-recapture cycle repeats with quantum and biological computing — and every historical precedent says it will — the concentration of computational power in the next era will dwarf anything this trilogy has documented. A quantum computer that can break encryption renders every current security system obsolete. A biological computer that can design molecules renders pharmaceutical development, materials science, and potentially biological weapons accessible to whoever controls the substrate. The entities that hold these capabilities will not be governments in the traditional sense. They will be the organizations — seven today, perhaps fewer tomorrow — that have the capital, the talent, and the infrastructure to build and operate these systems. The jar will be held by a very small number of hands. And the question of whether it is possible to take it back will have a different answer than it has had in every previous era.

Watch the substrates. Watch who controls access to quantum computing when it becomes operational at scale. Watch who funds biological computation research and who owns the results. Watch whether the governance frameworks being built today — the EU AI Act, the national strategies, the international declarations — address the current phase of AI or the phase that follows it. If they address only the current phase, then the governance will be obsolete before the substrate shift arrives, and the concentration will proceed without constraint. Again.

The Hegemons and the Hyperscalers

As this trilogy went to press in March 2026, the evidence was arriving from every direction simultaneously.

From the hegemons: the United States used AI in a military operation that killed eighty-three people in a sovereign nation, then threatened to blacklist the AI company that asked how its technology was used. It deployed economic coercion against an ally who named the rupture in the rules-based order, encouraged separatism within that ally's borders, and referred to the ally's elected leader as "governor." China expanded AI surveillance into minority language monitoring, manufactured deepfake narratives to counter evidence of repression, and exported its surveillance infrastructure to authoritarian governments in the Middle East and Central Asia. Russia poisoned Western AI training data with state-sponsored disinformation, deployed deepfakes across thirty-eight countries' elections, and used the information environment as a weapon with a sophistication that makes the Facebook/Myanmar case from 2017 look primitive.

From the hyperscalers: one company controls sixty percent of active satellites and has twice determined the trajectory of a major war through unilateral connectivity decisions. The cloud providers set the terms on which every downstream AI application operates. The model builders determine, through content policies and safety filters, what AI can and cannot do — governance decisions made without democratic mandate that affect billions of people. Open-source releases democratize capability while precluding downstream governance of that capability. And the AI companies that attempt to self-govern — to draw red lines, to maintain safety commitments — are threatened with blacklisting, contract cancellation, and replacement by competitors willing to provide unrestricted access.

The pattern is the same from every direction: power concentrates, the keeper is removed, and governance is absent. The hegemons use AI to project power without accountability. The hyperscalers accumulate infrastructure without democratic mandate. And the people who bear the consequences — the Venezuelan civilians, the Cuban families, the Uyghur communities, the defendants scored by algorithms, the workers displaced by automation, the nations whose sovereignty depends on infrastructure they do not own — have no seat at any table where the rules are being discussed.

> *"The pattern is the same from every direction: power concentrates, the keeper is removed, and governance is absent. The hegemons use AI to project power. The hyperscalers accumulate infrastructure. And the people who bear the consequences have no seat at any table."*

What's at Stake

I want to be direct about what I believe the evidence shows, because the trilogy's measured tone should not be mistaken for measured stakes.

If the scenario documented in *Who Holds the Jar?* unfolds without governance and without responsible keepers — if quantum and biological computing concentrate in a handful of entities without the frameworks that could distribute their benefits or constrain their power — then what follows is not a correction or a disruption or a transition. What follows is a permanent restructuring of who can compute, who can know, who can verify, and who can act. The entities that hold the jar will be able to solve problems the rest of the world cannot even formulate. They will be able to break encryption that the rest of the world depends on. They will be able to design molecules the rest of the world cannot analyze. They will be able to model systems the rest of the world cannot simulate. The asymmetry will not be a gap. It will be a wall.

For the select few who hold the jar, this is an era of extraordinary possibility. For everyone else, it is an era of extraordinary dependence — dependence on entities whose capabilities you cannot verify, whose decisions you cannot challenge, and whose interests may not align with yours. The keeper will not merely be removed from the loop. The keeper will not be able to understand the loop. The gap between those who compute and those who cannot will be not a difference of degree but a difference of kind — as fundamental as the difference between literacy and illiteracy in the age of the printing press, but operating at a scale and speed that makes the printing press's impact look gradual.

This is not a prediction. It is a trajectory. Every data point in this trilogy — every era of computation that concentrated power, every keeper who was removed, every governance framework that arrived too late — points in the same direction. The trajectory can be altered. The room can convene. The rules can be written. The keepers can be preserved. But the trajectory will not alter itself. It requires people who see it to name it, and people who name it to act, and people who act to do so before the window closes.

> *"For the select few who hold the jar, this is an era of extraordinary possibility. For everyone else, it is an era of extraordinary dependence. The keeper will not merely be removed from the loop. The keeper will not be able to understand the loop."*

A Final Word

I am a person who works in technology, not a policymaker, not a head of state, not a researcher with institutional authority. I wrote these books because the evidence demanded them and because nobody with more authority was writing them fast enough. The evidence may be incomplete. The analysis may be wrong in places. The predictions may not hold. I accept all of this in advance and welcome correction — correction is how the argument gets better, and an argument that matters deserves to be made as well as possible.

What I am not willing to accept is silence. The evidence is sufficient to act. The room must convene. The rules must be written. The keepers must be preserved. And the people who bear the consequences must have seats at the table where the rules are discussed — not because their inclusion will produce perfect governance, but because their exclusion will produce governance that serves only the people who are already present.

Imperfect governance, written now, by imperfect people, in an imperfect process. Exactly like Philadelphia. Exactly like Bretton Woods. Exactly like the NPT. And exactly like Mark Carney at Davos — a person who stood before the most powerful audience in the world and named the thing the room had been performing around, because someone had to speak first.

A Note on Method

This trilogy was written with the assistance of AI. Specifically, I used Anthropic's Claude as a research partner, drafting collaborator, fact-checker, and editorial reviewer throughout the writing process. I disclose this because the argument of these books demands it.

A trilogy that calls for transparency about how AI systems are used cannot exempt itself from that standard. Every chapter in *The Last Keeper* argues that the human must remain in the loop. I remained in the loop. Every factual claim was verified against primary sources. Every argument is mine. Every editorial decision — what to include, what to cut, what to emphasize, what to question — was a human judgment. The AI was a tool. The keeper was present.

I note, without irony, that during the writing of this trilogy, the AI company whose tool I used became the subject of one of the book's central case studies — the confrontation between Anthropic and the Pentagon over red lines in military AI. I did not plan this. The evidence arrived while the book was being written, which is itself a demonstration of how fast this field moves. I used the tool made by the company I was writing about, to write the book that argues for governing both the tool and the company. If that circularity troubles the reader, I understand. It troubles me too. I chose to name it rather than hide it.

The jar is filling.
The keeper is watching.
The room is waiting.
Speak.

Pumulo Sikaneta

Index

Page numbers refer to the complete trilogy edition.

A

B

C

D

E

F

G

H

I

K

L

M

N

O

P

R

S

T

U

V

W

X

www.ingramcontent.com/pod-product-compliance
Lightning Source LLC
Chambersburg PA
CBHW071445140726
47997CB00005B/1601